HVDC for Grid Services in Electric Power Systems

HVDC for Grid Services in Electric Power Systems

Special Issue Editor

Gilsoo Jang

MDPI • Basel • Beijing • Wuhan • Barcelona • Belgrade

Special Issue Editor
Gilsoo Jang
Korea University
Korea

Editorial Office
MDPI
St. Alban-Anlage 66
4052 Basel, Switzerland

This is a reprint of articles from the Special Issue published online in the open access journal *Applied Sciences* (ISSN 2076-3417) from 2018 to 2019 (available at: https://www.mdpi.com/journal/applsci/special_issues/HVDC)

For citation purposes, cite each article independently as indicated on the article page online and as indicated below:

LastName, A.A.; LastName, B.B.; LastName, C.C. Article Title. *Journal Name* **Year**, *Article Number*, Page Range.

ISBN 978-3-03921-762-5 (Pbk)
ISBN 978-3-03921-763-2 (PDF)

Contents

About the Special Issue Editor

Gilsoo Jang received his BE from the Department of Electrical Engineering, Korea University, Korea in 1991; MS degree from the Department of Electrical Engineering, Korea University, Korea in 1994; and Ph.D from the Department of Electrical Engineering, Iowa State University, USA in 1997. He was a visiting scientist at Department of Electrical & Computer Engineering, Iowa State University, USA from 1997 to 1998, a senior researcher at Power Systems Lab., Korea Electric Power Research Institute, Korea from 1998 to 2000, and a visiting professor at the Department of Electrical Engineering, Cornell University, USA from 2006 to 2007.

Dr. Jang has been a professor at the School of Electrical Engineering, Korea University since 2000. He worked as an Associate Dean of the College of Engineering from 2011 to 2103 and worked as a Dean of the School of Electrical Engineering from 2015 to 2017. With the support of the National Research Foundation of Korea since 2017, he has been the director of the Grid Connected Power Electronics Research Center.

He has participated in academic societies as a journal editor of *IEEE Transactions on Smart Grid*, as a technical reviewer for international journals and conferences, as a conference organizer, and so on. He is a senior member of IEEE and KIEE, and is a member of CIGRE.

Preface to "HVDC for Grid Services in Electric Power Systems"

Modern power systems are huge non-linear systems interconnected with many power generation and transmission systems. The large-scale deployment of variable renewable energy resources (RESs) has caused significant concerns about grid stability and power quality, and finding ways to control these large-scale systems is essential for stable operation. The control capabilities of HVDC and FACTS equipment can improve the dynamic behavior and flexibility of the grid, and various studies are underway in both system- and component-level modeling, control, and reliability. Unlike AC systems where the power flow of each branch is determined, the use of HVDC that can control power flow can be a solution to increase flexibility. However, if the set-points for control are selected incorrectly, it can have a negative impact on the power system. Therefore, in order to provide various services to the power system, it is important to operate an HVDC well at the appropriate location.

This Special Issue is intended to present a new HVDC topology or operation strategy to prevent abnormal grid operating conditions. The papers published in this Special Issue fall into several representative themes.

1. Improved HVDC operation and control strategies for system stability

-Analysis of Six Active Power Control Strategies of Interconnected Grids with VSC-HVDC

-A Virtual Impedance Control Strategy for Improving the Stability and Dynamic Performance of VSC–HVDC Operation in Bidirectional Power Flow Mode

-Novel Transient Power Control Schemes for BTB VSCs to Improve Angle Stability

-Development of a Loss-Minimization-Based Operation Strategy for Embedded BTB VSC HVDC

2. HVDC protection strategies in contingency cases

-A Novel Overcurrent Suppression Strategy during Reclosing Process of MMC-HVDC

-Assessment of Appropriate MMC Topology Considering DC Fault Handling Performance of Fault Protection Devices

-A Study on Stability Control of Grid Connected DC Distribution System Based on Second Order Generalized Integrator-Frequency Locked Loop (SOGI-FLL)

3. HVDC planning and implementation strategies, and field test experience

-A Frequency–Power Droop Coefficient Determination Method of Mixed Line-Commutated and Voltage-Sourced Converter Multi-Infeed, High-Voltage, Direct Current Systems: An Actual Case Study in Korea

-Application of a DC Distribution System in Korea: A Case Study of the LVDC Project

-A Quantitative Index to Evaluate the Commutation Failure Probability of LCC-HVDC with a Synchronous Condenser

We hope that these papers will help to achieve good results by applying HVDCs in power systems.

Gilsoo Jang
Special Issue Editor

applied sciences

MDPI

Editorial

Special Issue on HVDC for Grid Services in Electric Power Systems

Gilsoo Jang

School of Electrical Engineering, Korea University, 145 Anam-ro, Seongbuk-gu, Seoul 02841, Korea; gjang@korea.ac.kr; Tel.: +82-2-3290-3246

Received: 2 October 2019; Accepted: 7 October 2019; Published: 12 October 2019

1. Introduction

The modern electric power system has evolved into a huge nonlinear complex system, due to the interconnection of a lot of generation and transmission systems. The unparalleled growth of renewable energy resources (RES) has caused significant concern regarding grid stability and power quality, and it is essential to find ways to control such a massive system for effective operation. The controllability of HVDC and FACTs devices allows for improvement in the dynamic behavior and flexibility of grids. Research is being carried out at both the system and component levels of modelling, control, and stability. This Special Issue aims to present novel topologies or operation strategies to prevent abnormal grid conditions. The papers published in this special issue are categorized into several representative themes and briefly summarized the main contents of each paper.

2. Improved HVDC Operation and Control Strategy for System Stability

Analysis of Six Active Power Control Strategies of Interconnected Grids with VSC-HVDC [1]

In this paper, the generator angle stability of several active power control schemes of VSC-based HVDC is evaluated for interconnected two AC systems. Furthermore, in order to effectively evaluate angle stability, the Generators-VSC Interaction Factor index is newly implemented to distinguish the participating generators group which reacts to the converter power change.

A Virtual Impedance Control Strategy for Improving the Stability and Dynamic Performance of VSC–HVDC Operation in Bidirectional Power Flow Mode [2]

This paper adds the control loop to improve the performance and eliminate the steady-state error in the existing virtual impedance control of VSC HVDC. This paper eliminates the steady state with an additional control loop and verifies it by experiment and modeling.

Novel Transient Power Control Schemes for BTB VSCs to Improve Angle Stability [3]

This paper proposes two novel power control strategies to improve the angle stability of generators using a Back-to-Back (BTB) system-based voltage source converter. The power control strategy can emulate the behavior of the ac transmission to improve the angle stability while supporting the ac voltage at the primary level of the control structure.

Development of A Loss Minimization Based Operation Strategy for Embedded BTB VSC HVDC [4]

Using the power transfer distribution factor (PTDF), the HVDC-sensitive AC lines are classified as a monitoring line in advance, and a strategy for determining operating point for normal/emergency conditions is proposed.

3. HVDC Protection Strategy in Contingency Cases

A Novel Overcurrent Suppression Strategy during Reclosing Process of MMC-HVDC [5]

This paper discusses a strategy to control the overcurrent that can occur in the post-fault process of mesh current method (MMC)-HVDC. A MCM is proposed for accurate overcurrent calculation of a loop MMC-HVDC grid, and a reclosing current limiting resistance (RCLR) is calculated using the result.

Assessment of Appropriate MMC Topology Considering DC Fault Handling Performance of Fault Protection Devices [6]

This paper compared DC fault handling performance in variable fault location on a DC line. The simulation result confirmed that Half Bridge-MMC with a hybrid circuit breaker (HCB) is superior than FB-MMC with a residual circuit breaker (RCB) due to low fault current, low interruption time, low overvoltage magnitude and faster recovery.

A Study on Stability Control of Grid Connected DC Distribution System Based on Second Order Generalized Integrator-Frequency Locked Loop (SOGI-FLL) [7]

This paper proposes advanced control method using Second Order Generalized Integrator-Frequency Locked Loop (SOGI-FLL) that can be applied to a 3-phase AC/DC PWM converter for DC distribution. The proposed control scheme improves transient characteristics of DC distribution systems.

4. HVDC Planning & Implementation Strategy, and Field Test Experience

A Frequency–Power Droop Coefficient Determination Method of Mixed Line-Commutated and Voltage-Sourced Converter Multi-Infeed, High-Voltage, Direct Current Systems: An Actual Case Study in Korea [8]

In this paper, a new frequency-power droop coefficient determination method for a mixed line-commutated converter (LCC) and voltage-sourced converter (VSC)-based multi-infeed HVDC (MIDC) system is proposed. An interior-point method is used as an optimization algorithm to implement the proposed scheduling method, and the droop coefficients of the HVDCs are determined graphically using a Monte Carlo sampling method.

Application of a DC Distribution System in Korea: A Case Study of the LVDC Project [9]

This paper demonstrates DC distribution with real field test results. The authors also propose the operating procedures for an insulation monitoring device (IMD) and its algorithm. The real field test result in Gwangju was analyzed and the authors checked real IMD operation procedures.

A Quantitative Index to Evaluate the Commutation Failure Probability of LCC-HVDC with a Synchronous Condenser [10]

An index is proposed to allow quantitative evaluation of the positive effects on the commutation failure probability of LCC HVDC before and after the synchronous condenser is installed. This paper provides a rationale for the capacity allocation of synchronous condensers in LCC HVDC Projects.

Acknowledgments: This issue would not have been possible without the help of a variety of talented authors, professional reviewers, and the dedicated editorial team of Applied Sciences. Thank you to all the authors and reviewers for this opportunity. Finally, thanks to the Applied Sciences editorial team.

Conflicts of Interest: The author declares no conflict of interest.

References

1. Song, S.; Yoon, M.; Jang, G. Analysis of Six Active Power Control Strategies of Interconnected Grids with VSC-HVDC. *Appl. Sci.* **2019**, *9*, 183. [CrossRef]

2. Li, Y.; Liu, K.; Liao, X.; Zhu, S.; Huai, Q. A Virtual Impedance Control Strategy for Improving the Stability and Dynamic Performance of VSC–HVDC Operation in Bidirectional Power Flow Mode. *Appl. Sci.* **2019**, *9*, 3184. [CrossRef]
3. Song, S.; Hwang, S.; Ko, B.; Cha, S.; Jang, G. Novel Transient Power Control Schemes for BTB VSCs to Improve Angle Stability. *Appl. Sci.* **2018**, *8*, 1350. [CrossRef]
4. Lee, J.; Yoon, M.; Hwang, S.; Jeong, S.; Jung, S.; Jang, G. Development of A Loss Minimization Based Operation Strategy for Embedded BTB VSC HVDC. *Appl. Sci.* **2019**, *9*, 2234. [CrossRef]
5. Jiang, B.; Gong, Y. A Novel Overcurrent Suppression Strategy during Reclosing Process of MMC-HVDC. *Appl. Sci.* **2019**, *9*, 1737. [CrossRef]
6. Lee, H.-Y.; Asif, M.; Park, K.-H.; Lee, B.-W. Assessment of Appropriate MMC Topology Considering DC Fault Handling Performance of Fault Protection Devices. *Appl. Sci.* **2018**, *8*, 1834. [CrossRef]
7. Kang, J.-W.; Shin, K.-W.; Lee, H.; Kang, K.-M.; Kim, J.; Won, C.-Y. A Study on Stability Control of Grid Connected DC Distribution System Based on Second Order Generalized Integrator-Frequency Locked Loop (SOGI-FLL). *Appl. Sci.* **2018**, *8*, 1387. [CrossRef]
8. Lee, G.; Moon, S.; Hwang, P. A Frequency–Power Droop Coefficient Determination Method of Mixed Line-Commutated and Voltage-Sourced Converter Multi-Infeed, High-Voltage, Direct Current Systems: An Actual Case Study in Korea. *Appl. Sci.* **2019**, *9*, 606. [CrossRef]
9. Kim, J.; Kim, H.; Cho, Y.; Kim, H.; Cho, J. Application of a DC Distribution System in Korea: A Case Study of the LVDC Project. *Appl. Sci.* **2019**, *9*, 1074. [CrossRef]
10. Sha, J.; Guo, C.; Rehman, A.; Zhao, C. A Quantitative Index to Evaluate the Commutation Failure Probability of LCC-HVDC with a Synchronous Condenser. *Appl. Sci.* **2019**, *9*, 925. [CrossRef]

applied sciences

MDPI

Article

Analysis of Six Active Power Control Strategies of Interconnected Grids with VSC-HVDC

Sungyoon Song [1], Minhan Yoon [2] and Gilsoo Jang [1,*]

[1] School of Electrical Engineering, Korea University, Anam-ro, Sungbuk-gu, Seoul 02841, Korea; blue6947@korea.ac.kr

[2] Department of Electrical Engineering, Tongmyong University, Sinseon-ro, Nam-gu, Busan 48520, Korea; minhan.yoon@gmail.com

* Correspondence: gjang@korea.ac.kr; Tel.: +82-010-3412-2605

Received: 29 November 2018; Accepted: 31 December 2018; Published: 6 January 2019

Abstract: In this paper, the generator angle stability of several active power control schemes of a voltage-source converter (VSC)-based high-voltage DC (HVDC) is evaluated for two interconnected AC systems. Excluding frequency control, there has been no detailed analysis of interconnected grids depending upon the converter power control, so six different types of active power control of the VSC-HVDC are defined and analyzed in this paper. For each TSO (transmission system operator), the applicable schemes of two kinds of step control and four kinds of ramp-rate control with a droop characteristic are included in this research. Furthermore, in order to effectively evaluate the angle stability, the Generators-VSC Interaction Factor (GVIF) index is newly implemented to distinguish the participating generators (PGs) group which reacts to the converter power change. As a result, the transient stabilities of the two power systems are evaluated and the suitable active power control strategies are determined for two TSOs. Simulation studies are performed using the PSS®E program to analyze the power system transient stability and various active power control schemes of the VSC-HVDC. The results provide useful information indicating that the ramp-rate control shows a more stable characteristic than the step-control for interconnected grids; thus, a converter having a certain ramp-rate slope similar to that of the other generator shows more stable results in several cases.

Keywords: grid-interconnection; active power control strategies; transient stability; GVIF index; angle spread; VSC-HVDC

1. Introduction

Presently, renewable energy resources are considered a best practice in the response to global warming, and these power resources are concentrated in remote areas in order to effectively generate power. However, several instability issues arising from uneven large power generation requires TSOs (transmission system operators) to complement the grid structure [1]. Moreover, based on References [2–4], grid interconnection is emerging as an effective alternative for solving stability problems. For example, Nordic power systems in which several grids are interconnected by AC or DC links have increasingly accepted renewable energy resources, and have updated their hourly power exchange clauses [5]. This additional effort has led to the mitigation of several instability issues caused by uneven power generation, and many research works have also reported that renewable energy resources have become more accepted in many other countries [6].

In order to interconnect two different power systems, there are two options for TSOs: AC or DC lines. Nowadays, grid interconnection using an AC transmission line has a problem that increases the system complexity from the operation point of view, and may adversely decrease the system reliability. In fact, large blackouts have clearly confirmed that the close coupling of the neighboring systems might also include the risk of uncontrolled cascading effects in large and heavily loaded systems [7]. Furthermore, the AC system is vulnerable to sub-sea transmission connections and long interconnection; thus, the DC system, which has the advantage of high controllability, has been widely deployed for grid interconnection projects [8]. Considering the grid strength as the SCR (short circuit ratio) at each point, it is well known that the LCC (line commutated converter)-based high-voltage DC (HVDC) is restricted in that the converter cannot work properly if the connected AC system is weak [9]. Conservatively, in the case of AC systems with an SCR lower than 1.5, synchronous condensers have to be installed so as to increase the SCR of the AC system. In addition, the reactive power should be compensated depending upon the power sent, which reduces the simplicity of controllability in LCCs. The voltage-source converter (VSC)-HVDC has similar stability issues; however, it offers significant advantages such as high controllability, reliability, and small size. Benefiting from the significant technical advances in insulated gate bipolar transistors (IGBT), the VSC has become a competitive alternative to the LCC, so the VSC-HVDC is more commonly deployed nowadays. In the VSC system, two main stability issues have generally been presented in detail to date:

(1) Operation region of the VSC-HVDC

The reactive power of the VSC-HVDC can be limited according to the AC grid voltage and the equivalent impedance. In addition, the DC voltage control and PLL (Phase-Locked Loop) can restrict the power angle to approximately 51° for a stable operation without the support of the dynamic reactive power [10]. In order to obtain an improved power transfer capability from the VSC-HVDC, the X/R ratio (the ratio of the system reactance to the system resistance) and the impedance angle must be considered. Therefore, the SCR index representing the grid strength is an important factor from the perspective of the capability region.

(2) Dynamic performance of the VSC-HVDC

In previous studies on the relationship between the PLL and the VSC-HVDC, many authors have mentioned that a converter with large PLL gains that is connected to a weak AC grid ($2 < SCR$) is prone to instabilities when subjected to a disturbance [11]. This is because the PLL that is used for the angle-reference generation can easily generate an unstable eigenvalue with high proportional gains. The AC voltage phase is highly sensitive to the d and q current injections of the converter in a weak grid. Detailed results have been described in a few references [12,13].

However, the stability issues mentioned above can be resolved by the robust compensator design mentioned by many authors [12–15]. The robust PI (Proportional and Integral) parameters, feedforward controllers, and adjusted PLL parameters enable the stable operation of VSC-HVDC, and the damping condition which occurs at a certain frequency range can be mitigated. Therefore, in this work, the VSC-HVDC system is deployed without considering the small signal stability issues, and the main contribution of this paper is to analyze the impact of six active power control strategies on the generator angle stability of two interconnected grids.

Excluding the contingency event, the fixed power control for two grids is commonly used to lessen the operation burdens of TSOs. However, during a contingency event, the power should be adjusted to provide grid services such as frequency support or transient stability support. According to the previous works related to VSC control for grid service, Reference [16] demonstrated the AC transmission emulation control strategy, which acts like an AC line when a contingency event occurs. It is able to mitigate the possible overloading of adjacent AC transmission, and maintain power balance between metropolitan regions. However, the transferred power is not exactly estimated since the output power is varied depending upon contingency event types; thus, it is not suitable for interconnected grids since there is a clear exchange clause in their agreement. In Reference [17], the flexible operation of the generator tripping scheme was achieved without a large decelerating energy as the generators trip, and it was confirmed that a simple converter control strategy that transfers the maximum power reserve instantly to the fault area surely contributes to the stability of the AC network. However, this paper only addressed one kind of step control. In References [18,19], the DC voltage droop, local frequency control, and weighted-average frequency control are compared in detail; however, this analysis was performed in an embedded MTDC (multi-terminal DC) environment. In References [20,21], the kinds of frequency–power modulation control strategies for the converter to enhance the system transient stability are introduced. As can be observed, there has been no detailed analysis of interconnected grids depending upon the several converter power control schemes in a point-to-point environment. Therefore, in this paper, two step control and four ramp-rate control scenarios are specifically defined, and then simulated to provide useful feasibility studies results for grid operators.

A dynamic control model of the VSC-HVDC is developed written by Fortran language in the PSS®E program (Power Transmission System Planning Software), and the ideal averaged equivalent VSC model is used. The MMC (modular multilevel converter) is not used since the AC system stability is the major observation target. To perform this analysis, the GVIF index, meaning the Generators-VSC Interaction Factor, is newly defined in Section 2. In Section 3, the introduction of the VSC-HVDC model serves to illustrate the configuration of the active power controller. In addition, six active power control strategies are defined in Section 4. Lastly, a simulation of the transient stability regarding the control schemes is performed.

2. Identify PGs (Participating Generators) with the GVIF Index

As shown in Figure 1, areas 1 and 2 are interconnected by a VSC-HVDC link. The initial DC power direction is from area 1 to area 2, so the converter is used to provide auxiliary service for area 2. This grid structure may be in the form of an interconnection link between countries or between regions [18]. If multiple generators are connected in parallel to each area, it is difficult to detect which generator contributes to incremental power according to the converter power change. The TSOs must determine which generators respond to the converter control, and this process is needed to distinguish the participating generators (PGs) group.

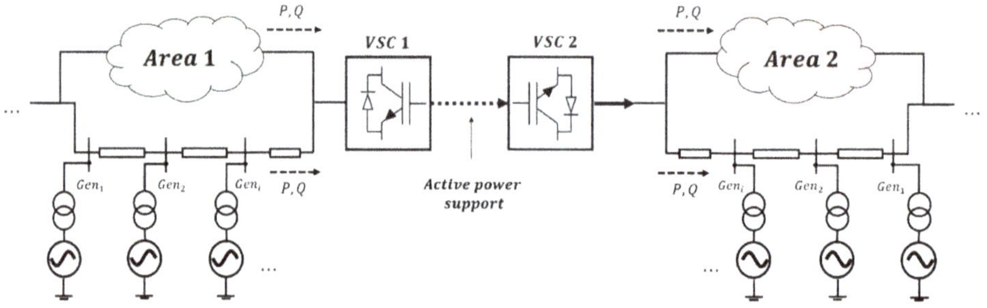

Figure 1. Two interconnected grids with a voltage-source converter (VSC)-based high-voltage DC (HVDC).

A traditional synchronous generator consists of a governor and a prime mover to support frequency regulation. The simplified first-order differential equation of the dynamic generator model is shown in Equations (1) and (2), where P_v is the valve position of the governor; P_{ref} is the power reference of governor; R is the droop value; P_m is the prime mover output power; and T_G and T_P are the time constants of the governor and prime mover, respectively [22].

$$\frac{d\Delta P_v}{dt} = -\frac{\Delta P_v}{T_G} + \frac{\Delta P_{ref}}{T_G} - \frac{1}{T_G R}\Delta f. \tag{1}$$

$$\frac{d\Delta P_m}{dt} = -\frac{\Delta P_m}{T_P} + \frac{\Delta P_v}{T_P}. \tag{2}$$

Based on Equations (1) and (2), the generators react to grid frequency change as Δf. If all generators have the same frequency droop value, an individual generator increases in the same MW power in a liner decrease in speed corresponding to the percent droop selected and no-load frequency. However, in the real grid operation, the frequency measurement result as Δf is slightly different at each region at the same time; thus, the ΔP_v and ΔP_m were made unlike the expected values. During the dynamic state, the generator output power is mainly determined by the droop value as R, but it is also related to the distance to the point at which the frequency change occurs. As a result, the approximate incremental output of each generator with the droop slope can be estimated, but it is difficult to derive the exact incremental power from each generator. In order to consider both the governor droop value and the electrical distance between the generators and the converter, the new grouping index, referred to as the Generators-VSC Interaction Factor (GVIF) to select the PGs is implemented as follows.

$$GVIF_{i,j} = \frac{\Delta P_i}{\Delta P_j}. \tag{3}$$

where bus i is the generator bus connected in area 1. Bus j is the VSC-HVDC bus, and as we know, the multi-infeed HVDC system has several bus positions as $j \geq 2$. The GVIF is the dynamic active power change of bus j over the active power change of bus i. When the active power change ΔP_j is 1%, the active power change ratio of bus i is the GVIF. If the generators have the same frequency–power droop value, the electrical distance is the main factor impacting the GVIF since the frequency measurement results are slightly different at each region. Thus, using the GVIF, the frequency measurement result errors could be corrected on each generator output. In the steady-state condition, since the frequency change point is always the converter bus as bus j, the generators with a high GVIF index could be considered to be closer to the converter or to have a high droop value. The generator which has a zero value of GVIF does not participate in the incremental power generation. In this paper, we define the generators with GVIF > 0 as PGs, and the angle stabilities of all PGs are evaluated by the general transient stability index as angle spread in Section 5.

3. Active Power Controller Design of VSC-HVDC

The schematic diagram of the VSC-HVDC is illustrated in Figure 2. The widely used vector controller is applied in the VSC. Let the converter side impedance be simply modeled as a series-connected three phase inductor and resistor, and the AC grid in the *abc* frame is:

$$\begin{bmatrix} v_1^d \\ v_1^q \end{bmatrix} - \begin{bmatrix} v_2^d \\ v_2^q \end{bmatrix} = R\begin{bmatrix} i_d \\ i_q \end{bmatrix} + L\frac{d}{dt}\begin{bmatrix} i_d \\ i_q \end{bmatrix} - \begin{bmatrix} -\omega L i_q \\ \omega L i_d \end{bmatrix}, \tag{4}$$

where the v_2 is the voltage at PCC and v_1 is the voltage at the converter. In addition, R and L are the resistance and inductance, respectively, and i is the current flowing to the AC grid. Filter components prevent the generation of harmonic current by the converter, and they also affect the stability between the AC grid and the VSC.

The reference voltage generated by the inner current control loop is transformed back into the *abc* frame and used for Pulse With Modulation (PWM) to produce the desired converter three-phase voltage. The voltage reference sent to the PWM is represented by:

$$\begin{bmatrix} \Delta v_2^d \\ \Delta v_2^q \end{bmatrix} = -\begin{bmatrix} A_d(s) \\ A_q(s) \end{bmatrix}\begin{bmatrix} \Delta i_{d,ref} - \Delta i_d \\ \Delta i_{q,ref} - \Delta i_q \end{bmatrix} + \omega L\begin{bmatrix} -\Delta i_q \\ \Delta i_d \end{bmatrix} + \begin{bmatrix} v_1^d \\ v_1^q \end{bmatrix}, \tag{5}$$

where $A_d(s)$ *and* $A_q(s) = \frac{k_p s + k_i}{s}$.

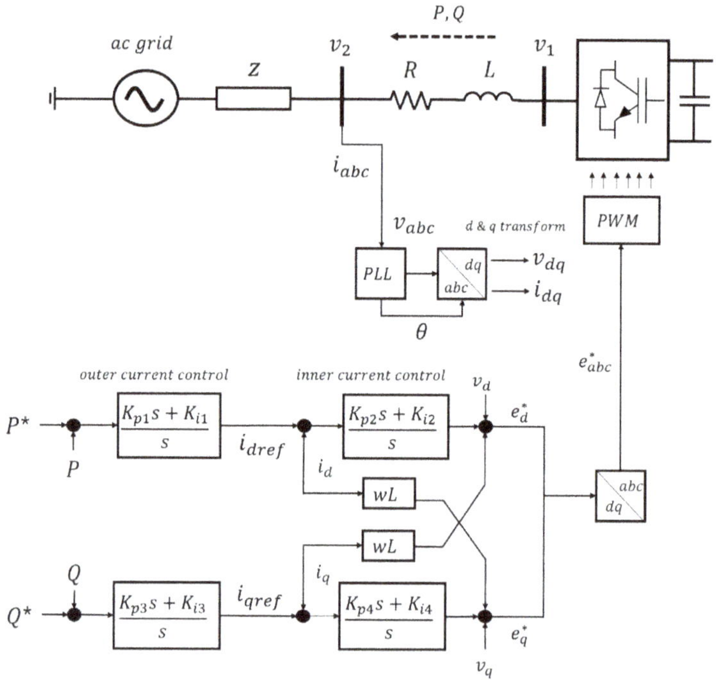

Figure 2. VSC control block diagram. PWM: Pulse With Modulation; PLL Phase-Locked Loop.

The PWM switching delay is then approximated by a first-order Padé approximation as follows:

$$G_{PWM}(s) = \frac{1 - \frac{1.5T_d s}{2}}{1 + \frac{1.5T_d s}{2}}, \tag{6}$$

where T_d is the switching delay in the PWM. Then, combining (5) and (6), the equation can be rearranged by the input terms as v_2^{dq} and $i_{dq,ref}$, with i_{dq} as the output. The transfer functions of the inner controller are expressed by:

$$i_d = A \cdot v_2^d + B \cdot i_{d,ref}, \tag{7}$$

$$i_q = A \cdot v_2^q + B \cdot i_{q,ref}, \tag{8}$$

where $A = \frac{1 - G_{PWM}(s)}{(R+Ls)+G_{PWM}(s) \cdot A_d(s)}$, $B = \frac{G_{PWM}(s) \cdot A(s)}{(R+Ls)+G_{PWM}(s) \cdot A_q(s)}$.

The q-axis current of the *d*-*q* frame is aligned with the AC system phasor based on the PLL, i.e., $i_q = 0$. Thus, the converter admittance is derived as i_{abc}/v_2, which is obtained as follows:

$$Y_{VSC}(s) = \frac{1 - G_{PWM}(s)}{(R + Ls) + G_{PWM}(s) \cdot A_d(s)}. \tag{9}$$

Depending on (9), the d-axis current flowing to the AC grid to control active power is represented in Figure 3. Following the block diagram, the stability between the AC grid and converter can be analyzed based on the initial operating point of the VSC. However, the detailed small signal analysis is not of interest in this paper, and the useful results were given with an impedance-based stability analysis theorem by References [23,24].

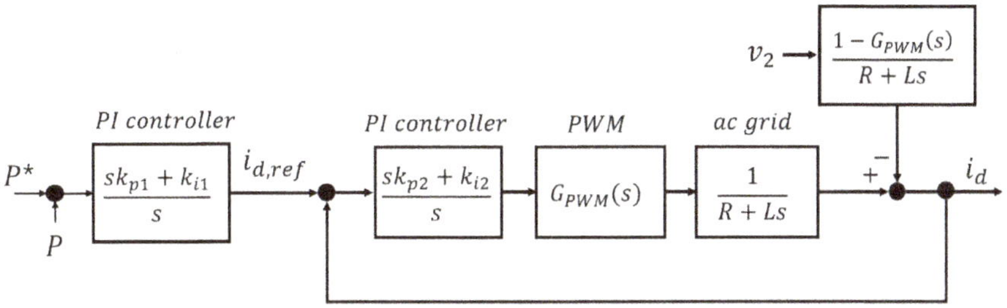

Figure 3. D-axis current control structure of the VSC; PI: Proportional and Integral.

Following the d-axis current response of the converter, the active power of the PGs is simultaneously activated with their own GVIF index. The incremental power of the PGs of area 1 is transferred to area 2, and its characteristic is adjusted by several active power control strategies, as illustrated in the following section.

4. Analysis of Active Power Control Strategies

Excluding the frequency control, two major active power control strategies can be applied for the VSC-HVDC. The first one is the step control, which releases active power step-by-step at certain times, as shown in Figure 4a. The second one is the ramp-rate control, which transfers active power with a specific slope, as illustrated in Figure 4b. In this section, each control strategy is introduced and then defined.

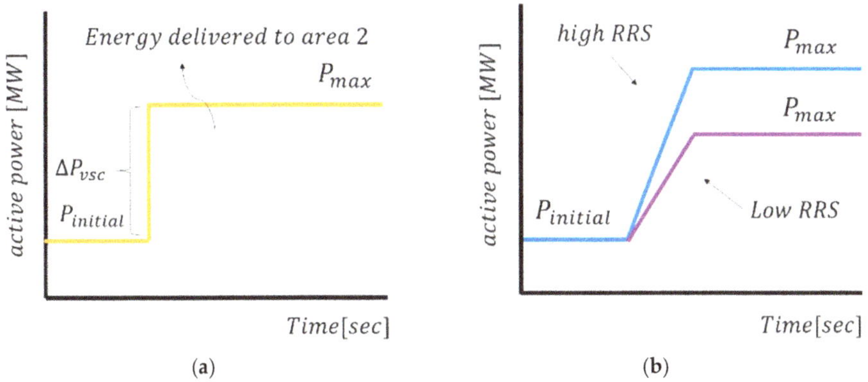

Figure 4. Two major active power control schemes: (**a**) step control (**b**) ramp-rate control with different ramp-rate slopes (RRSs).

4.1. Introduction of Step Control and Ramp-Rate Control Strategies

As is generally known, the step control sustains its initial power in normal stable operation, then increases power at certain times, when area 2 has a frequency drop or emergency event. By contrast, the ramp-rate control has a preset ramp-rate slope (RRS), as shown in Figure 5. The power changes from one stable state to another stable state with a ramping event, and considering a discrete time representation, the ramp-rate of P_{vsc} at the kth instant can be determined using the following expression:

$$RRS = \frac{dP_{vsc}}{dt}(k) = \frac{[P_{vsc}(k) - P_{vsc}(k-1)]}{t(k) - t(k-1)}. \tag{10}$$

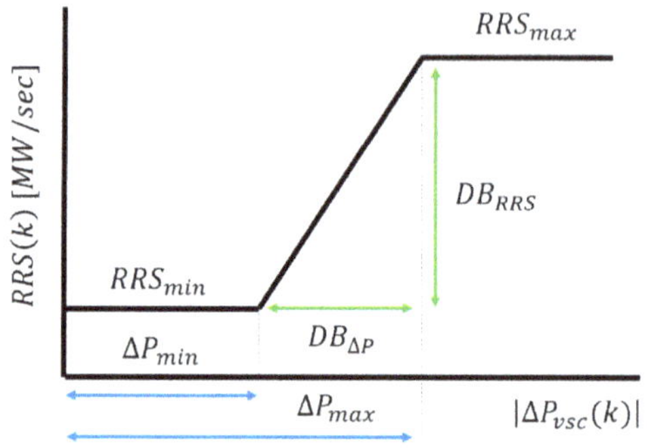

Figure 5. Different RRSs of ramp-rate control; DB: difference between.

The RRS can vary depending upon $|\Delta P_{vsc}(k)|$, which is the absolute value of the deviation between $P_{vsc}(k-1)$ and the active power reference as P_{vsc}^{ref}. If the $|\Delta P_{vsc}(k)|$ is large, a high RRS is applied so as to ensure a fast response. On the other hand, if the power order as P_{vsc}^{ref} is small, a low RRS is applied to follow the final P_{vsc}^{ref} value. Such a droop characteristic is given in Equation (11) with Figure 5.

$$RRS(k) = \begin{cases} RRS_{min},\ if\ |\Delta P_{vsc}(k)| < \Delta P_{min} \\ RRS_{min} + \frac{DB_{RRS}}{DB_{\Delta P}} \times [|\Delta P_{vsc}(k)| - \Delta P_{min}], \\ if\ \Delta P_{min} \leq |\Delta P_{vsc}(k)| \leq \Delta P_{max} \\ RRS_{max},\ if\ |\Delta P_{vsc}(k)| > \Delta P_{max} \end{cases}. \tag{11}$$

The RRS(k) is a droop-based desired ramp-rate; RRS_{max} and RRS_{min} are the max and min ramp-rate slopes of ΔP_{vsc}, respectively; DB_{RRS} is the difference between and RRS_{max} and RRS_{min}; ΔP_{min} and ΔP_{max} are the lower and upper bands of dynamic change of active power variation; and $DB_{\Delta P}$ is the difference between and ΔP_{min} and ΔP_{max}. As mentioned previously, if the converter receives a high P_{vsc}^{ref} order by the TSO, the active power sharply increases with a ramp rate of RRS_{max}, and a large active power is transferred from area 1 to area 2. Based on References [16,17], it was confirmed that converter power control is helpful when the power system has a contingency event; therefore, this control strategy largely contributes to the frequency stability of area 2. However, we can also expect that the angle stability of area 1 could be further worsened. On the other hand, if the converter reaches RRS_{min} when the $|\Delta P_{vsc}(k)|$ is smaller than ΔP_{min}, a small amount of active power is delivered to area 2, and the angle stability of the PGs in area 1 will be more improved than that in the RRS_{max} case.

4.2. More Detailed Description of Step Control and Ramp-Rate Control Strategies

There is an N-1 contingency event in area 2 with the given scenario, which is represented in Figure 6. Each country, if it has a different Special Protection System (SPS), as the generator tripping schemes are generally called, commonly commands the specific generators to be tripped so as to balance the network power. Basically, the nine cycles as time delay should be taken with the generator tripping scheme, since a mechanical switch is included. As the electronic power equipment only requires communication delay, the activating time is naturally fast [17]. In this paper, three cycles of communication delay are adopted for t_1 time, and the minimum frequency occurred at t_2 time, as shown in Figures 6 and 7. Accordingly, various active power control strategies can be applicable to the VSC-HVDC to supply more power to area 2. Given the frequency fluctuation of area 2, six active power control strategies in total are introduced hereafter.

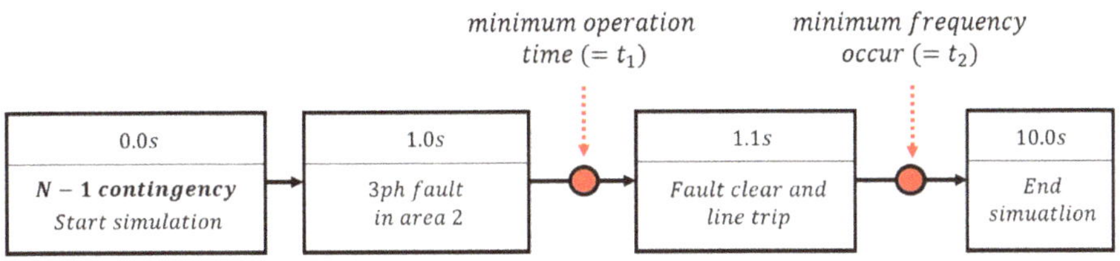

Figure 6. Contingency scenario in area 2.

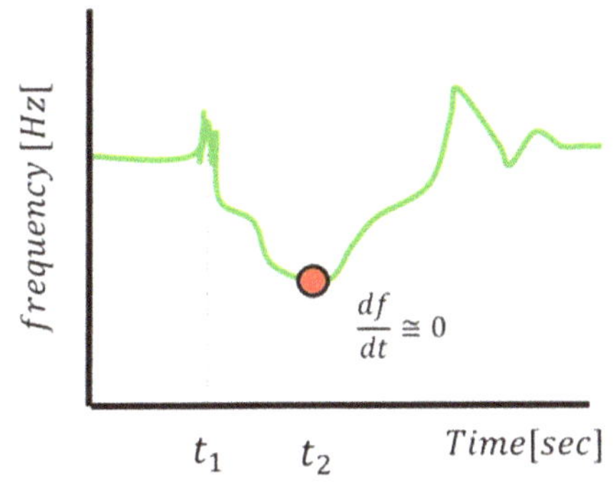

Figure 7. Frequency drop during a contingency event in area 2.

Figure 8a,b both show step control, with the only difference being a power release time. In (a), the power increased at t_1, which is the minimum time to receive an SPS signal as 1.05 s. In (b), the converter changes its power at t_2. Benefiting from the fast response of power electronics, (a) and (b) show advantages of being able to control the power step-by-step, unlike the common generator characteristics. Except for (a) and (b), all other control strategies are ramp-rate control strategies with different RRSs and sending times. Figure 8c,d have different RSSs between t_1 and t_2, which are derived from two different power references. More specifically, the power command in (c) is twice the value of (d); thus, the RRS of (d) is selected to be half. This is to observe the results according to both initial power support speed and amount. Figure 8e,f discuss how the angle stability of each area changed during the recovery stage of area 2. Thus, the two control strategies have the same power reference value, but different RSSs between t_2 and t_3. For this, the t_3 is selected for the control variable to adjust the RSS. In the simulation study, the t_3 of (e) is twice the value of (f), so that (e) has an RRS twice that of (f). According to the control characteristic, the six control schemes are defined in Table 1, and specific simulation results are introduced hereafter.

Table 1. Six different active power control strategies.

Denomination	Control Type	RRS	Control Time
(a)	Step	-	t_1
(b)	Step	-	t_2
(c)	Ramp-rate	RRS_{max}	t_1 to t_2
(d)	Ramp-rate	RRS_{min}	t_1 to t_2
(e)	Ramp-rate	RRS_{max}	t_2 to t_3
(f)	Ramp-rate	RRS_{min}	t_2 to t_3

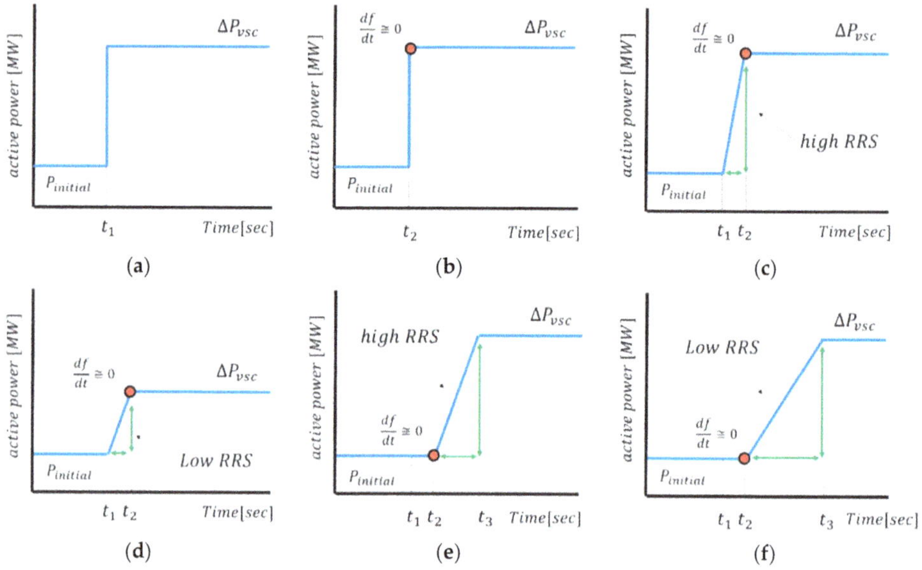

Figure 8. Six active power control schemes: (**a**) step control at t_1; (**b**) step control at t_2; (**c**) ramp-rate control with high RRS between t_1 and t_2; (**d**) ramp-rate control with low RRS between t_1 and t_2; (**e**) ramp-rate control with high RRS between t_2 and t_3; (**f**) ramp-rate control with low RRS between t_2 and t_3.

5. Transient Stability Simulation Results with Six Different Active Power Control Schemes

The interconnected grid configuration for simulation studies is represented in Figure 9, and the three-phase fault event at 345 kV AC transmission occurs in area 2. The detailed system parameters are defined in Table 2. The averaged equivalent circuit of the two-level VSC is used for this feasibility study, and it was supplemented by one on-state switch resistance in each phase, and an equivalent current source at the DC side. Note that the VSC model should be injected by current sources in the PSS®E program, and the outer and inner current controller parameters are adjusted to achieve the desired system response. Accordingly, the six different power output characteristics of the VSC-HVDC are simulated, as shown in Figure 10. There are two power control start times of t_1 and t_2. (a), (c), and (d) supply power at t_1, and (b), (e), and (f) control power at t_2. The RRS and power reference values are all different according to each control characteristic.

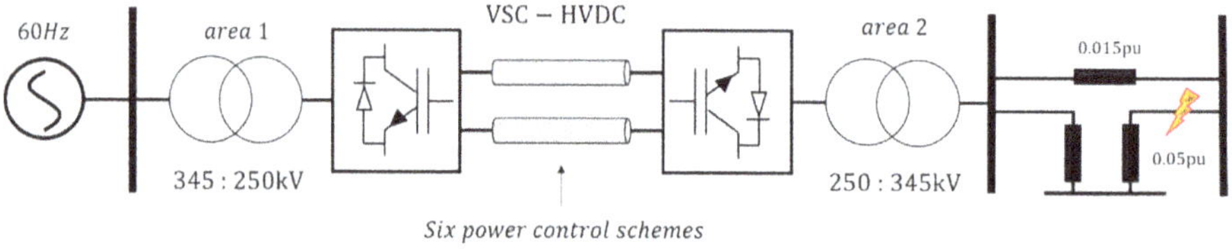

Figure 9. System configuration for simulation studies.

Table 2. System parameters.

Grid Parameters	Value	Converter Parameters	Value
Grid frequency	60 Hz	VSC HVDC Rated capacity	1200 MVA
Total load	94,463 MW	VSC HVDC DC link voltage	250 kV
Total generation	95,802 MW	DC capacitance	1500 µF
AC voltage of each side	345 kV	k_{p1}, k_{p3}	0.5
Short circuit ratio	25	k_{p2}	0.65
Leakage reactance	0.07 pu	k_{p4}	0.8
Transformer Voltage ratio	345/250 250/345	k_{i1}, k_{i3}, k_{i4}	0.01
Transformer rating	1200 MVA	k_{i2}	0.1

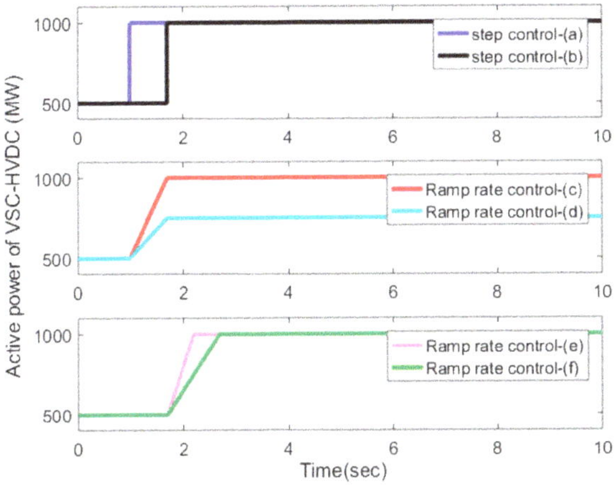

Figure 10. Six power control schemes of the VSC-HVDC.

As mentioned in Section 2, in order to select PGs when the converter controls its power, the GVIF index of the Korean power system is analyzed. The PGs are selected using Equation (3) with ΔP_{vsc} = 100 MW, and the PGs list is given in Table 3. As a total of 269 generators are connected in the Korean grid at the peak load condition; however, for the sake of convenience, only 22 generators with GVIF are illustrated in this paper.

Table 3. Generators-VSC Interaction Factor (GVIF) analysis of participating generators (PGs) in area 1 ($\Delta P_{vsc} = 100$ MW).

Generator Number	Generator List	Initial Power (MW)	GVIF	PGs (O: PG, X: Not PG)
1	Yunghung #G1	800	0.007	O
2	Yunghung #G2	800	0.007	O
3	Yunghung #G3	870	0.01	O
4	Yunghung #G4	870	0.01	O
5	Yunghung #G5	870	0.01	O
6	Yunghung #G6	870	0.01	O
7	Sininchen #G9	167.7	0.004	O
8	Sinin #GT10	167.7	0.004	O
9	Sinin #GT11	168.34	0.003	O
10	Sinin #GT12	168.34	0.003	O

Table 3. *Cont.*

Generator Number	Generator List	Initial Power (MW)	GVIF	PGs (O: PG, X: Not PG)
11	Sininchen #S9	181.26	0	X
12	Inchen #GT1	164.065	0.004	O
13	Inchen #GT3	180.69	0.003	O
14	Inchen #GT5	168	0.005	O
15	Inchen #ST1	163.02	0	X
16	Inchen #ST2	178.12	0	X
17	Inchen #ST3	167.1	0	X
18	POS5 #GT1	210.52	0.005	O
19	POS5#GT2	210.52	0.005	O
20	POS6 #GT3	217.17	0.005	O
21	POS6 #GT4	224.4	0.005	O
22	POS6 #ST2	232.65	0	X
23		⋮		

As can be observed in Table 3, the steam turbine (ST) has a zero value of GVIF, since it has a zero droop value, so the active power is not adjusted in response to the frequency variation. On the other hand, the gas turbine has a certain droop value that is more sensitively activated than the steam turbine during the dynamic converter power control. In order to correct the different frequency measurement results, the GVIF index is used to select PGs. We conclude that the power transmitted to area 2 comes from the generators with GVIF values larger than zero. In addition, as shown in Table 3, the most influential PGs are *Yunghung #G3~#G6* with GVIF = 0.01. With the driven PGs list, the angle spread, which is the difference between the largest and smallest participating machine angles, is analyzed for interconnected areas.

5.1. Angle Spread Evaluation with PGs

In order to evaluate angle spread, the Overcorrections index, which is generally used for control response determination, is introduced in this paper. Assume that $x_{initial}$ is the original steady state value of variable x, x_F is the first encountered peak of x during a transient event, and x_T is the second encountered peak of x during a transient event.

$$x_0 = \left|\frac{x_{initial} - x_F}{x_{initial}}\right| + \left|\frac{x_{initial} - x_T}{x_{initial}}\right| \tag{12}$$

x is defined in this paper as the angle spread. Thus, the larger x_0 is, the worse the angle stability of the PGs, and if x_0 is large enough, there is a possibility of loss of synchronisms. The locus of x_0 as well as the brief conclusions are well illustrated with each scheme hereafter.

5.1.1. Step Control

The results in Figure 11 show the angle spread of two interconnected grids with the control scheme as (a) and (b). Both control schemes are step control strategies, and we aimed to determine whether the active power should be transmitted at t_1 or t_2. In area 1, the only difference between the two schemes is the power release time, so we may see only the result of area 2. From the angle spread results of area 2, it can be seen that sending power at t_2 can further improve the angle stability. More specifically, the first damping of angle spread as x_F after the contingency event is mitigated at about 1.45 s, and the recovery characteristic is more improved from 2.1 s to 3 s. This is because the power from area 1 contributes to the frequency recovery characteristic of area 2. However, in (a), the power is transmitted immediately after the fault, so it contributes less to the angle stability of area 2. The reason for this is that the initial frequency drop driven by fault is momentarily increased due to the immediate active power support. Therefore, the PGs of area 2 generate a relatively small amount of power according to the small frequency drop. However, this process finally makes a slow frequency recovery, and the angle stability of the PGs becomes worse than that in (b). Therefore, we can conclude that sending power during the frequency recovery stage is more beneficial to both TSOs.

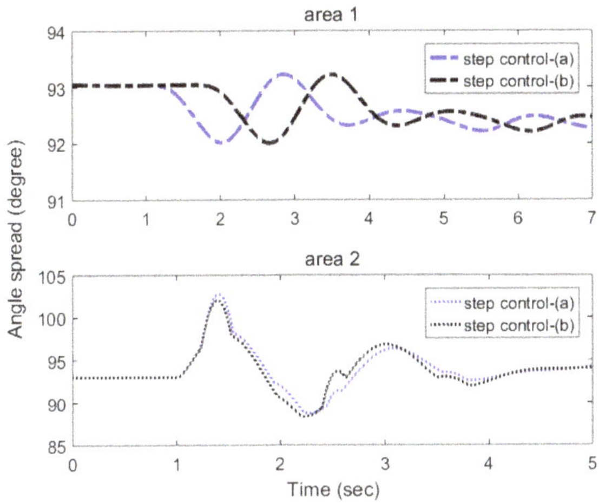

Figure 11. Angle spread results of each area according to (a) and (b) in Table 1.

5.1.2. Ramp-Rate Control

Figure 12 shows the results regarding (c) and (d). Both control schemes are ramp-rate control strategies, and both the RRS between t_1 and t_2 and the active power reference are different. In (c), 1000 MW power is transferred to area 2 with a high RRS, so the angle stability of the sending side is further worsened compared to that in (d), which sends 750 MW. In area 1, (c) experiences a sudden large power change at t_1, and shows a more unstable result. Concerning the receiving side in area 2, the angle deviation x_0 is explicitly different between the two methods. The results show that the generators of area 2 could not cover the severe frequency drop, and there is a possibility of loss of synchronisms at time = 1.45 s. Therefore, applying method (c) is more stable in terms of the angle stability of area 2. However, a tradeoff was observed as method (d) is more suitable for area 1, where the SCR is low.

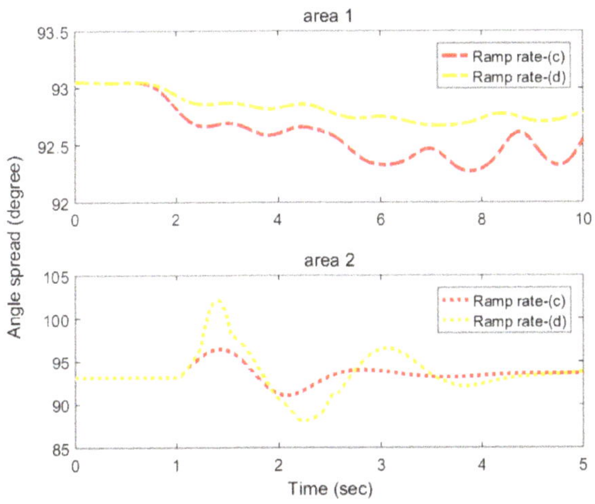

Figure 12. Angle spread results of each area according to (c) and (d) in Table 1.

Figure 13 shows the results regarding (e) and (f). Both control schemes are ramp-rate control strategies, and the only difference is the RRS between t_2 and t_3. The power change references are all the same as 1000 MW; thus, the $|\Delta P_{vsc}(k)|$ is also the same. By adjusting time t_3, the two control methods were set to have a twofold slope difference. Scheme (f), which has a low RRS, shows slightly greater angle stability results in area 1. From the result of area 2, scheme (e) shows a greater angle stability result that the first encountered peak time and its recovery characteristic is faster than that in (f). Thus, it is concluded that active power control with a certain RRS is more suitable for both areas. However, using an overly low RRS will limit the stability improvement of area 2.

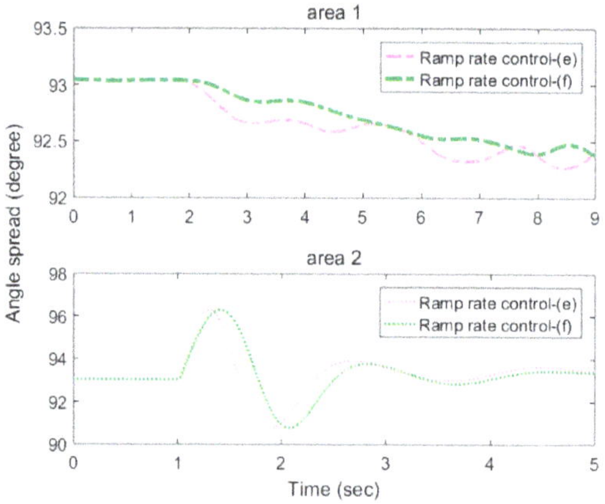

Figure 13. Angle spread results of each area according to (e) and (f) in Table 1.

Figure 14 shows the results regarding (c) and (e). Both control schemes are ramp-rate control strategies with the same RRS, and the only difference is the power sending time. In (c), the power is transmitted immediately following the contingency event (=t_1) while the power is increased during the frequency recovery stage (=t_2) in (e). As shown in the result of area 2, if the grid operator uses both ramp-rate control schemes, sending power during the frequency recovery stage shows more stable results in terms of the angle stability. This is because, as mentioned earlier in Section 5.1.1, the initial frequency drop will not be immediately reduced by (c). This characteristic causes the PGs of area 2 to generate less power. Therefore, at about 1.3 s the first damping angle x_F is more mitigated with (e); thus, sending power at t_2 is more suitable for both TSOs.

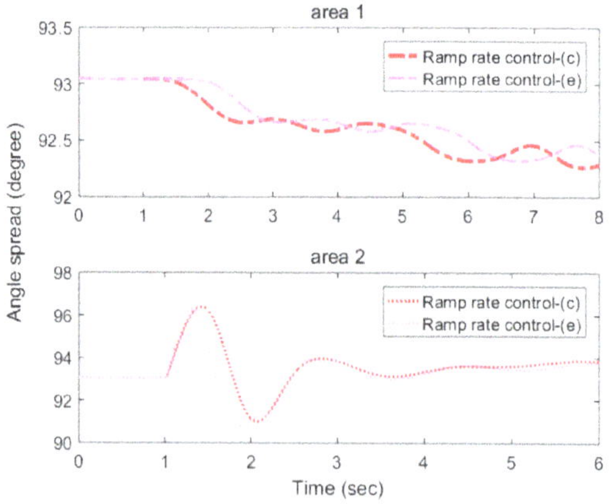

Figure 14. Angle spread results of each area according to (c) and (e) in Table 1.

5.1.3. Step Control vs. Ramp-Rate Control

The results in Figures 15 and 16 are about (a) and (e), and (b) and (e), respectively. The main purpose of this simulation is to compare the step control and ramp-rate control strategies; thus, this comparison is the main result of this paper. The results explicitly show that the angle stability with the step control is more unstable in both areas. Using scheme (e) has more smaller value of x_0 than (a) as shown in Table 4. As a result, the VSC-HVDC connecting two different grids should have its own RRS considering both the sending and receiving side generator stability. Thus, it is recommended that the two TSOs should include the ramp-rate control clauses in the HVDC design phase when the interconnected mutual agreement is introduced.

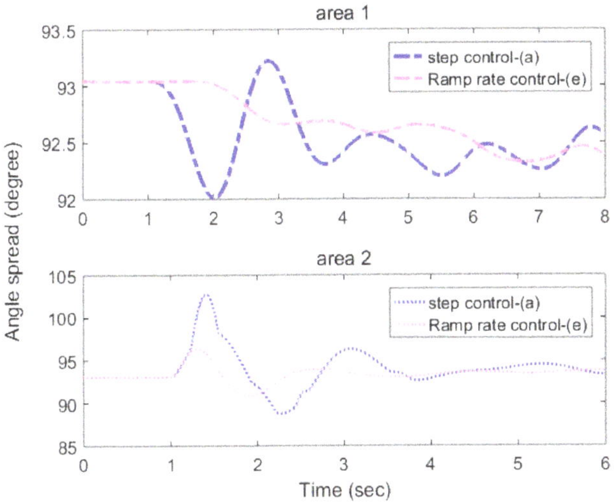

Figure 15. Angle spread results of each area according to (b) and (e).

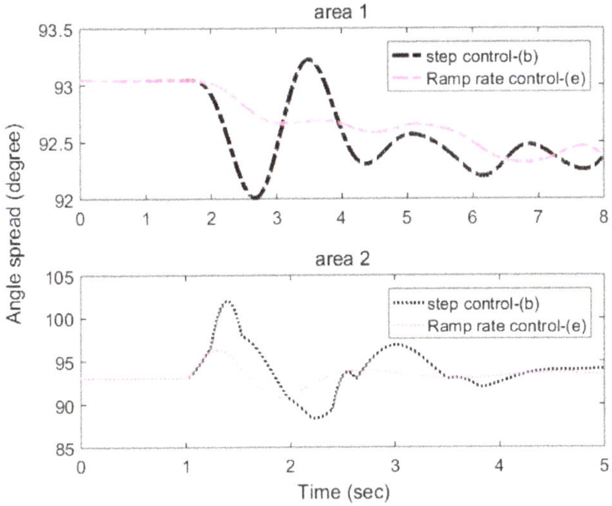

Figure 16. Angle spread results of each area according to (a) and (e).

Table 4. Overcorrection results of (a) and (e) in area 2.

Target System	Control Strategy	$x_{initial}$	x_F	x_T	x_0
area 2	(a)	93.04	102.42	88.6	0.147
	(e)		96.21	90.82	0.05

5.2. Detailed Comparison of Six Power Control Strategies

The suitable control strategy was recommended for each TSO as shown in Table 5. It is confirmed that sending the active power at time t_2 is more effective than that at t_1. This is because the reserve power is helpful during the frequency recovery stage. Furthermore, the ramp-rate control is more suitable than the step control to both areas since the sudden large power change causes the large angle variation of PGs. In addition, the mismatch of active power could further worsen the angle stability, so if the required active power is not estimated, the converter should have its own RRS which is not too far from the generators' output characteristic. Furthermore, if the N-1 or N-2 contingency event occurs at the sending side while the power is transmitted to the receiving side, the ramp-rate control will show a more stable result at the sending side grid. At last, the TSOs should consider the SCR of two grids based on the driven results, and select the appropriate scheme as shown in Table 5. This information will be useful for two TSOs planning grid interconnection projects.

Table 5. Suitable control scheme for each TSO (O: Most suitable/Δ: Suitable/X: Not suitable). SCR: short circuit ratio.

Target System	(a)	(b)	(c)	(d)	(e)	(f)
Receiving side with high SCR	Δ	Δ	O	O	O	O
Receiving side with Low SCR	X	X	O	Δ	O	Δ
Sending side with high SCR	Δ	Δ	O	O	O	O
Sending side with Low SCR	X	X	Δ	O	O	O

6. Conclusions

To date, the frequency control for interconnected grids has been analyzed in several works. However, using frequency control increases the operation burden of one side. Thus, the impact of various types of active power control scheme should be simulated and analyzed. Furthermore, if there is a power increase clause between interconnected grids, this issue will be more critical for grid operators. In this paper, when the receiving side has an emergency event, six applicable active power control strategies are defined and simulated.

Before the simulation, our researchers expected that the step control, which is a powerful strength of power electronics, could further improve the angle stability of PGs due to its fast response. Contrary to expectations, the results show that the VSC-HVDC with certain RRSs similar to the other generator output characteristics provides more stable results in several cases. Our findings also confirm that sending power during the frequency recovery stage is more effective for the receiving side, whereas sending power right after the fault degrades the generator output characteristic since the low initial frequency error is measured at the generator side. To prevent this problem, one solution can be applied. If the droop slope of the generator is adjusted when the converter starts the power control, the small initial frequency error caused by the converter support can be corrected at each governor. This increases the generators output, and thus the frequency can reach the nominal value.

Furthermore, grid operators have to alter their control strategy based on the SCR of each grid base, as shown in Table 5. These results provide useful information to grid operators, and the advantages and disadvantages of each control scheme are well shown.

Author Contributions: The main idea was proposed by S.S. and G.J.; the experiment results were collected and analyzed by S.S. and M.Y.

Funding: This research received no external funding.

Acknowledgments: This work was supported under the framework of international cooperation program managed by National Research Foundation of Korea (No. 2017K1A4A3013579) and also supported by "Human Resources Program in Energy Technology" of the Korea Institute of Energy Technology Evaluation and Planning (KETEP), granted financial resource from the Ministry of Trade, Industry and Energy, Korea. (No. 20174030201540).

Conflicts of Interest: The authors declare no conflict of interest.

Nomenclature

θ	Angle difference between $\dot{E}_1$ and $\dot{E}_2$
$\dot{E}_1$	Internal generators voltage
$\dot{E}_2$	Infinite bus voltage
w	Rotor speed
H	Shaft inertia
P_i	Mechanical power
P_n	Electrical power
v_1	voltage at voltage source converter
v_2	voltage at PCC
R	Resistance between AC grid and converter
L	Inductance between AC grid and converter
i	Ac current from converter to AC grid
k_p, k_i	PI controller in current controller of VSC
T_d	Switching delay in PWM
RRS	Ramp-rate slope of VSC
$\lvert\Delta P_{vsc}(k)\rvert$	Absolute value of the deviation between $P_{vsc}(k-1)$ and active power reference as P_{vsc}^{ref}
x_0	Overcorrections for the evaluation of angle spread

References

1. Yang, D.; Wang, X.; Liu, F.; Xin, K.; Liu, Y.; Blaabjerg, F. Adaptive reactive power control of PV power plants for improved power transfer capability under ultra-weak grid conditions. *IEEE Trans. Smart Grid* **2017**. [CrossRef]
2. Purvins, A.; Wilkening, H.; Fulli, G.; Tzimas, E.; Celli, G.; Mocci, S.; Pilo, F.; Tedde, S. A European supergrid for renewable energy: Local impacts and far-reaching challenges. *J. Clean. Prod.* **2011**, *19*, 1909–1916. [CrossRef]
3. Singh, M.; Khadkikar, V.; Chandra, A.; Varma, R.K. Grid interconnection of renewable energy sources at the distribution level with power-quality improvement features. *IEEE Trans. Power Deliv.* **2011**, *26*, 307–315. [CrossRef]
4. Feltes, J.; Gemmell, B.; Retzmann, D. From smart grid to super grid: Solutions with HVDC and FACTS for grid access of renewable energy sources. In Proceedings of the 2011 IEEE Power and Energy Society General Meeting, Detroit, MI, USA, 24–29 July 2011; pp. 1–6.
5. Entsoe. Agreement Regarding Operation of the Interconnected Nordic Power System (System Operation Agreement). 2006. Available online: https://docstore.entsoe.eu/Documents/Publications/SOC/Nordic/System_Operation_Agreement_2014.pdf (accessed on 1 January 2019).
6. Aljohani, T.M.; Alzahrani, A.M. The Operation of the GCCIA HVDC Project and Its Potential Impacts on the Electric Power Systems of the Region. *Int. J. Electron. Electr. Eng.* **2014**, *2*, 207–213. [CrossRef]
7. Zhang, L.; Harnefors, L.; Nee, H.-P. Interconnection of two very weak AC systems by VSC-HVDC links using power-synchronization control. *IEEE Trans. Power Syst.* **2011**, *26*, 344–355. [CrossRef]
8. Flourentzou, N.; Agelidis, V.G.; Demetriades, G.D. VSC-based HVDC power transmission systems: An overview. *IEEE Trans. Power Electron.* **2009**, *24*, 592–602. [CrossRef]
9. Daryabak, M.; Filizadeh, S.; Jatskevich, J.; Davoudi, A.; Saeedifard, M.; Sood, V.; Martinez, J.; Aliprantis, D.; Cano, J.; Mehrizi-Sani, A. Modeling of LCC-HVDC systems using dynamic phasors. *IEEE Trans. Power Deliv.* **2014**, *29*, 1989–1998. [CrossRef]

10. Huang, Y.; Wang, D. Effect of Control-Loops Interactions on Power Stability Limits of VSC Integrated to AC System. *IEEE Trans. Power Deliv.* **2018**. [CrossRef]
11. Zhou, J.Z.; Ding, H.; Fan, S.; Zhang, Y.; Gole, A.M. Impact of Short-Circuit Ratio and Phase-Locked-Loop Parameters on the Small-Signal Behavior of a VSC-HVDC Converter. *IEEE Trans. Power Deliv.* **2014**, *29*, 2287–2296. [CrossRef]
12. Huang, Y.; Yuan, X.; Hu, J.; Zhou, P. Modeling of VSC connected to weak grid for stability analysis of DC-link voltage control. *IEEE J. Emerg. Sel. Top. Power Electron.* **2015**, *3*, 1193–1204. [CrossRef]
13. Davari, M.; Mohamed, Y.A.-R.I. Robust vector control of a very weak-grid-connected voltage-source converter considering the phase-locked loop dynamics. *IEEE Trans. Power Electron.* **2017**, *32*, 977–994. [CrossRef]
14. Wu, G.; Liang, J.; Zhou, X.; Li, Y.; Egea-Alvarez, A.; Li, G.; Peng, H.; Zhang, X. Analysis and design of vector control for VSC-HVDC connected to weak grids. *CSEE J. Power Energy Syst.* **2017**, *3*, 115–124. [CrossRef]
15. Suul, J.A.; D'Arco, S.; Rodríguez, P.; Molinas, M. Impedance-compensated grid synchronisation for extending the stability range of weak grids with voltage source converters. *IET Gener. Transm. Distrib.* **2016**, *10*, 1315–1326. [CrossRef]
16. Song, S.; Kim, J.; Lee, J.; Jang, G. AC Transmission Emulation Control Strategies for the BTB VSC HVDC System in the Metropolitan Area of Seoul. *Energies* **2017**, *10*, 1143. [CrossRef]
17. Song, S.; Hwang, S.; Ko, B.; Cha, S.; Jang, G. Novel Transient Power Control Schemes for BTB VSCs to Improve Angle Stability. *Appl. Sci.* **2018**, *8*, 1350. [CrossRef]
18. Chaudhuri, N.R.; Chaudhuri, B. Adaptive droop control for effective power sharing in multi-terminal DC (MTDC) grids. *IEEE Trans. Power Syst.* **2013**, *28*, 21–29. [CrossRef]
19. Renedo, J.; Garcia-Cerrada, A.; Rouco, L. Active power control strategies for transient stability enhancement of AC/DC grids with VSC-HVDC multi-terminal systems. *IEEE Trans. Power Syst.* **2016**, *31*, 4595–4604. [CrossRef]
20. Du, C.; Bollen, M.H. Power-frequency control for VSC-HVDC during island operation. In Proceedings of the 8th IEE International Conference on AC-DC Power Transmission, London, UK, 28–31 March 2006.
21. Alamuti, M.M.; Saunders, C.S.; Taylor, G.A. A novel VSC HVDC active power control strategy to improve AC system stability. In Proceedings of the PES General Meeting | Conference & Exposition, National Harbor, MD, USA, 27–31 July 2014; pp. 1–5.
22. Yang, T.; Zhang, Y.; Wang, Z.; Pen, H. Secondary Frequency Stochastic Optimal Control in Independent Microgrids with Virtual Synchronous Generator-Controlled Energy Storage Systems. *Energies* **2018**, *11*, 2388. [CrossRef]
23. Bayo-Salas, A.; Beerten, J.; Rimez, J.; Van Hertem, D. Impedance-based stability assessment of parallel VSC HVDC grid connections. In Proceedings of the 11th IET International Conference on AC and DC Power Transmission, Birmingham, UK, 10–12 February 2015.
24. Wen, B.; Dong, D.; Boroyevich, D.; Burgos, R.; Mattavelli, P.; Shen, Z. Impedance-based analysis of grid-synchronization stability for three-phase paralleled converters. *IEEE Trans. Power Electron.* **2016**, *31*, 26–38. [CrossRef]

applied sciences

MDPI

Article

A Virtual Impedance Control Strategy for Improving the Stability and Dynamic Performance of VSC–HVDC Operation in Bidirectional Power Flow Mode

Yuye Li *, Kaipei Liu, Xiaobing Liao, Shu Zhu and Qing Huai

School of Electrical Engineering and Automation, Wuhan University, Wuhan 430072, China
* Correspondence: liyuye@whu.edu.cn; Tel.: +86-134-7627-0643

Received: 19 July 2019; Accepted: 1 August 2019; Published: 5 August 2019

Abstract: It is a common practice that one converter controls DC voltage and the other controls power in two-terminal voltage source converter (VSC)–based high voltage DC (HVDC) systems for AC gird interconnection. The maximum transmission power from a DC-voltage-controlled converter to a power-controlled converter is less than that of the opposite transmission direction. In order to increase the transmission power from a DC-voltage-controlled converter to a power-controlled converter, an improved virtual impedance control strategy is proposed in this paper. Based on the proposed control strategy, the DC impedance model of the VSC–HVDC system is built, including the output impedance of two converters and DC cable impedance. The stability of the system with an improved virtual impedance control is analyzed in Nyquist stability criterion. The proposed control strategy can improve the transmission capacity of the system by changing the DC output impedance of the DC voltage-controlled converter. The effectiveness of the proposed control strategy is verified by simulation. The simulation results show that the proposed control strategy has better dynamic performance than traditional control strategies.

Keywords: VSC–HVDC; DC-side oscillation; virtual impedance; impedance-based Nyquist stability criterion

1. Introduction

With the development of power electronic devices, VSC–HVDC systems have been widely applied to AC grid interconnection because of their independent decoupling control of active and reactive power [1–3]. Recently, a large number of studies on modeling, control, and stability analysis of VSC–HVDC system have been published [4–10]. Previous studies have shown that the interaction between converters or between the converter and the grid influences the stability of a system. DC- side oscillation is a problem in VSC–HVDC and has been reported in a real project [11]. When DC side oscillation occurs, a DC system will not work and will impact on the power system. Therefore, the DC-side stability of VSC–HVDC should be evaluated before connecting it to the main grid.

In VSC–HVDC systems applied to AC grid interconnection, active power often needs bidirectional transmission [12]. However, studies show that the maximum transmission power of the DC-voltage-controlled converter to power-controlled converter is less than that in the opposite power flow direction [13]. An Impedance-based approach can be adopted to analyze the influence of VSC–HVDC systems with different directions of transmission power on stability [13].

The impedance stability criterion was proposed in [14] and used in grid-connected inverters [15]. It was applied as a stability criterion in a VSC–HVDC system [12,16–19]. The impedance model of two-terminal VSC–HVDCs was built in [12], and the cause of DC current resonance was analyzed with

Nyquist stability criteria. Different subsystems were selected to analyze the DC-side stability of the VSC–HVDC system with a transfer function method in [16]. The influence of overhead transmission lines, DC cables, and the DC-side filter on system stability was investigated in [17]. The different performances of the lumped parameter and distributed parameter circuit models in stability analysis were discussed in [18]. It was found that the distributed parameter circuit model is more accurate in stability analysis.

In the condition of VSC–HVDC for AC Grid interconnection, the maximum transmission capacities of the different power flow directions are different. Thus, a control strategy is required to improve the transmission capacity from the DC-voltage-controlled converter to the power-controlled converter to improve resource utilization efficiency. To increase the maximum transmission power, the DC side oscillation must be suppressed. The suppression methods of DC side oscillation can be classified into passive methods [20,21] and active methods [22–29]. Passive methods suppress resonance by introducing a passive damper branch into the circuit to remodel the impedance of the source converter or load converter in a cascaded system. Active methods suppress resonance by introducing voltage and current feedback control in a controller to improve the impedance of the source converter or load converter. Virtual impedance is widely used in control systems as an active damping control method [24–29]. It can be introduced to suppress DC-side oscillation [24,25], to limit output current for voltage controlled inverters during overloads or faults [26,27], to improve the stability of a grid-connected inverter by change its input admittance [28], and to enhance the small-signal stability of a modular multilevel converter (MMC) based DC grid [29].

Virtual impedance in the DC voltage control loop can suppress the DC-side oscillation of a VSC–HVDC transmission system and improve its stability margin and the transmission capacity of the system [24,25]. However, virtual impedance control leads to steady-state errors in the DC output voltage of DC-voltage-controlled converters, due to different operating points [25]. An improved virtual impedance control strategy is proposed in this paper. To design an appropriate control strategy, a stability analysis is required. Thus, a DC impedance-based model of VSC–HVDC is built, and the stability of the system is analyzed based on impedance stability criteria.

The rest of the paper is organized as follows: Section 2 describes the system's structure and the simulation of a VSC–HVDC system, as well as the design of an improved virtual impedance control. Section 3 presents the impedance model of an HVDC system. Section 4 conducts a stability analysis on the basis of impedance stability criteria, and Section 5 shows the simulation verification. Section 6 summarizes the proposed method.

2. Improved Virtual Impedance Control Principle

This paper mainly studies a two-level topology structure. The analysis in this paper can be also applicable to an MMC system if the dc bus voltage ripples are insignificant [12]. A two-level VSC–HVDC system used in AC grid interconnection is depicted in Figure 1. Figure 1 shows a DC voltage-controlled converter and a power-controlled converter on the left and right sides, respectively. The two converters have an identical structure. $R_{n1} + jX_{n1}$ and $R_{n2} + jX_{n2}$ are the equivalent impedance of the AC system, $R_{c1} + j\omega_1 L_1$ and $R_{c2} + j\omega_2 L_{c2}$ are the impedance of the filter reactor, C_{f1} and C_{f2} are the filter capacitors, $\dot{m}_1$ and $\dot{m}_2$ are the modulations of the converter station, $\dot{u}_{s1}$ and $\dot{u}_{s2}$ are the AC voltage at point of common coupling, $\dot{u}_{g1}$ and $\dot{u}_{g2}$ are the AC voltage of the AC system, $\dot{i}_{s1}$ and $\dot{i}_{s2}$ are the AC current flowing through the filter reactor, and $\dot{i}_{g1}$ and $\dot{i}_{g2}$ are the AC current flowing through the AC system. The DC cable is a π type, with an equivalent resistance of R_d, an equivalent inductance of L_d, and an equivalent capacitance of C_d.

The modeling and control of the VSC–HVDC system are presented in a synchronous rotating frame (SRF). The transformation of the three-phase quantity from stationary reference frame to the SRF is based on the amplitude-invariant Park transformation, with the d-axis aligned with the voltage vector u_s and q-axis leading the d-axis by 90°. The grid voltage defines the system's dq reference. A phase-locked loop (PLL) defines the controller *dq* reference. The system reference is aligned with the PLL reference in a steady state. When a small disturbance occurs, the system reference is no longer aligned with the PLL reference. The relationship between the system reference and the PLL reference under a small disturbance is shown in Figure 2.

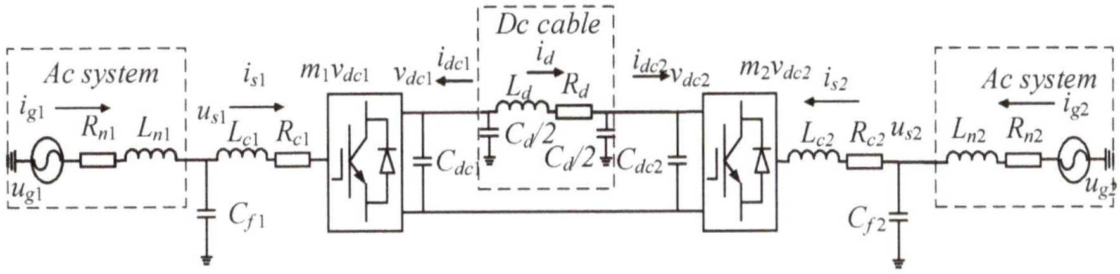

Figure 1. VSC–HVDC used in AC grid interconnection.

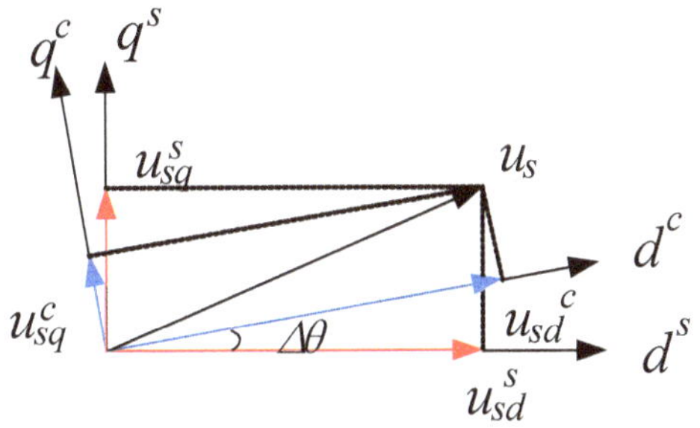

Figure 2. System and phase-locked loop (PLL) references.

Here, subscript *dq* represent the components of the physical quantity in the SRF. Subscript 0 represents the value of the physical quantity at the static working point, subscript *ref* represents the given value of physical quantity, and subscript Δ represents the small disturbance components of the physical quantity. Superscript *c* represents the components of physical quantity in the PLL reference, and superscript *s* represents the components in the system reference.

The simulation model of VSC–HVDC in Figure 1 is built on a MATLAB/Simulink. The system parameters are shown in the following table. The vector current control [13] is adopted in the two converters, and the control is shown in Figure 3. For symmetry of the VSC–HVDC system, subscripts 1,2 of the system parameters in Table 1 and the control diagram in Figure 3 are omitted. In the model, controller parameters are per unit. The base angular grid frequency is 50 Hz, the base grid voltage is 110 kV, the base system capacity is 500 MW, and the base DC voltage is 250 kV. According to the parameters in Table 1, the system's closed-loop bandwidth with a DC voltage controller is 65 Hz, and the system's open-loop phase margin is 58 degrees. The system's closed-loop bandwidth with an active power controller is 8 Hz, and the system's open-loop phase margin is 150 degrees. The system's closed-loop bandwidth with the current controller is 180 Hz, and the system's open-loop phase margin is 85 degrees.

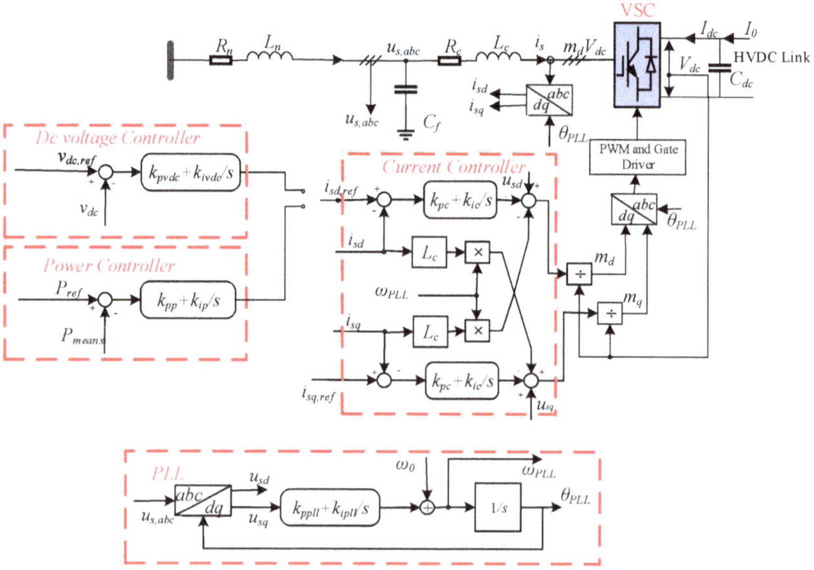

Figure 3. Control structure of the converters.

Table 1. Simulation parameters.

Parameters		Values
Converter and AC system	System capacity Sn/MW	500
	Line voltage of grid u_g/kV	110
	DC voltage V_{dc}/kV	250
	Grid internal resistance R_n/Ω	0.2
	Grid frequency ω_0/Hz	50
	Grid internal inductance L_n/H	1×10^{-3}
	Filter reactor inductance L_c/H	4.5×10^{-2}
	Filter reactor resistance R_c/Ω	0.2
	DC side capacitance $C_{dc}/\mu f$	300
DC cable	DC cable resistance R_d/Ω/km	1.39×10^{-2}
	DC cable inductance L_d/H/km	1.59×10^{-4}
	DC cable capacitance C_d/F/km	2.31×10^{-7}
Controller	DC voltage outer loop k_{pvdc}/k_{ivdc}	15/100
	Current inner loop k_{pc}/k_{ic}	0.5/0.1
	Active power outer loop k_{pp}/k_{ip}	1/10
	Phase locked loop k_{pPLL}/k_{iPLL}	10/100

The power flow direction from the power-controlled converter to the DC voltage-controlled converter is set to positive. Figure 4 shows the resulting time-domain responses of the DC voltage of the VSC–HVDC system. At 2 s, the active power instruction value steps from 500 MW to −500 MW, and the length of the DC cable is 50 km. It can be observed that the DC voltage starts to oscillate, the active power starts to fluctuate, and the system loses stability.

The power-controlled converter exhibits a constant power load (CPL) when the active power is transmitted from the DC-voltage controlled converter to the power-controlled converter. The incremental input resistance characteristic caused by CPL affects the stability of the VSC–HVDC system [24].

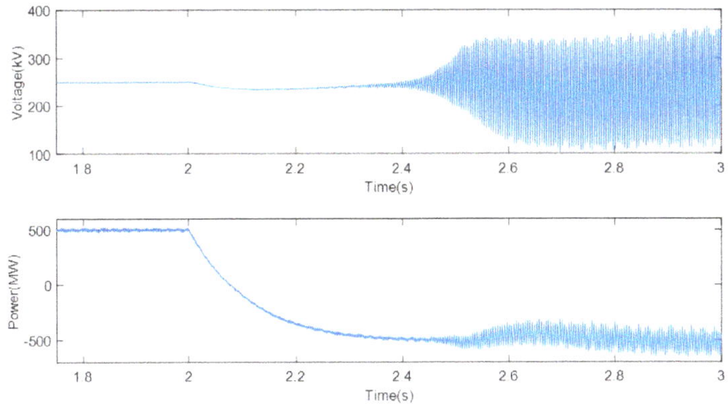

Figure 4. DC-link voltage and the active power of the VSC–HVDC system.

A virtual impedance control strategy is introduced in the DC-voltage-controlled converter to mitigate the DC-side oscillation caused by the negative incremental input resistance characteristic of the power-controlled converter. The expression of the virtual impedance control is

$$v_{dcref} = v_{dcn} - (R_{eq} + sL_{eq})i_{dc} \tag{1}$$

where v_{dcref} is the DC voltage reference value, v_{dcn} is the no-load DC voltage of the converter, and R_{eq} and L_{eq} are the set values for the virtual impedance. Due to the addition of current feedback, under loaded conditions, there will be a fixed steady-state error between the DC voltage reference value and the measured value, which almost equals $R_{eq} * i_{dc}$. Thus, the steady-state error increases with an increase in the transmission power (in both power flow directions).

In order to eliminate the steady-state error caused by virtual impedance, this paper modifies the voltage control loop, as shown by the blue dotted line frame in Figure 5.

$$i_{dcn} = \frac{k_i}{s}(v_{dcn} - v_{dc}) \tag{2}$$

$$v_{dcref} = v_{dcn} - (R_{eq} + sL_{eq})(i_{dc} - i_{dcn}) \tag{3}$$

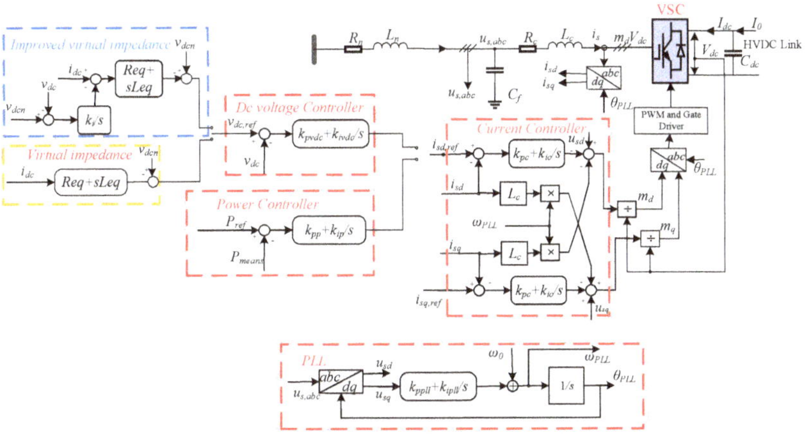

Figure 5. Proposed control structure of the converters.

3. Impedance Model of Converters with Improved Virtual Impedance Control Strategy

3.1. DC-Side Impedance Modeling of DC-Voltage-Controlled Converter

The linearized dynamic equations of the DC voltage-controlled converter are expressed as

$$\begin{bmatrix} m_{d0} \\ m_{q0} \end{bmatrix} \Delta v_{dc} + v_{dc0} \begin{bmatrix} \Delta m_d^s \\ \Delta m_q^s \end{bmatrix} + Z_0(s) \begin{bmatrix} \Delta i_{sd}^s \\ \Delta i_{sq}^s \end{bmatrix} = \begin{bmatrix} \Delta u_{sd}^s \\ \Delta u_{sq}^s \end{bmatrix} \tag{4}$$

$$\begin{bmatrix} \Delta u_{sd}^s \\ \Delta u_{sq}^s \end{bmatrix} = -Z_g(s) \begin{bmatrix} \Delta i_{gd}^s \\ \Delta i_{gq}^s \end{bmatrix} \tag{5}$$

$$\begin{bmatrix} \Delta i_{gd}^s \\ \Delta i_{gq}^s \end{bmatrix} = \begin{bmatrix} \Delta i_{sd}^s \\ \Delta i_{sq}^s \end{bmatrix} + Y_{cf}(s) \begin{bmatrix} \Delta u_{sd}^s \\ \Delta u_{sq}^s \end{bmatrix} \tag{6}$$

where

$$Z_0 = \begin{bmatrix} R_c + sL_c & -\omega_0 L_c \\ \omega_0 L_c & R_c + sL_c \end{bmatrix} \tag{7}$$

$$Z_g = \begin{bmatrix} R_n + sL_n & -\omega_0 L_n \\ \omega_0 L_n & R_n + sL_n \end{bmatrix} \tag{8}$$

$$Y_{cf}(s) = \begin{bmatrix} sC_f & -\omega_0 C_f \\ \omega_0 C_f & sC_f \end{bmatrix} \tag{9}$$

Its controller and modulator linearized equations are expressed as

$$\begin{bmatrix} \Delta i_{sd,ref} \\ \Delta i_{sq,ref} \end{bmatrix} = \begin{bmatrix} k_{pvdc} + \frac{k_{ivdc}}{s} \\ 0 \end{bmatrix} (\Delta v_{dcref} - \Delta v_{dc}) \tag{10}$$

$$v_{dc0} \begin{bmatrix} \Delta m_d^c \\ \Delta m_q^c \end{bmatrix} = -G_{pwm} G_{cc} \begin{bmatrix} \Delta i_{sd,ref} \\ \Delta i_{sq,ref} \end{bmatrix} + G_{pwm} \begin{bmatrix} \Delta u_{sd}^c \\ \Delta u_{sq}^c \end{bmatrix} + G_{pwm}(G_{cc} + Z_{del}) \begin{bmatrix} \Delta i_{sd}^c \\ \Delta i_{sq}^c \end{bmatrix} \tag{11}$$

where k_{pvdc} and k_{ivdc} are the proportional and integral gains of the DC voltage controller, respectively, G_{pwm} is the PWM delay, and G_{cc} is the current compensator transfer function, where k_{pc} and k_{ic} are the proportional and integral gains of the current compensator, respectively.

$$G_{pwm} = \begin{bmatrix} H_{pwm} & 0 \\ 0 & H_{pwm} \end{bmatrix} \tag{12}$$

$$H_{pwm} = e^{-sT_s} \frac{1 - e^{-sT_s}}{sT_s} \tag{13}$$

$$G_{cc} = \begin{bmatrix} k_{pc} + \frac{k_{ic}}{s} & 0 \\ 0 & k_{pc} + \frac{k_{ic}}{s} \end{bmatrix} \tag{14}$$

$$Z_{del} = \begin{bmatrix} 0 & \omega_{pLL} L_c \\ -\omega_{pLL} L_c & 0 \end{bmatrix} \tag{15}$$

The variables of the controller are based on the output of the PLL reference, whereas the variables of the circuit are based on the system reference. The relationship of physical quantity between the PLL and the system references is expressed as [13]

$$\begin{bmatrix} \Delta i_{sd}^c \\ \Delta i_{sq}^c \end{bmatrix} = \begin{bmatrix} \Delta i_{sd}^s \\ \Delta i_{sq}^s \end{bmatrix} + \overbrace{\begin{bmatrix} 0 & G_{PLL}(s) i_{sq0} \\ 0 & -G_{PLL}(s) i_{sd0} \end{bmatrix}}^{G_{PLL}^i} \begin{bmatrix} \Delta u_{sd}^s \\ \Delta u_{sq}^s \end{bmatrix} \tag{16}$$

$$\begin{bmatrix} \Delta u^c_{sd} \\ \Delta u^c_{sq} \end{bmatrix} = \overbrace{\begin{bmatrix} 1 & G_{PLL}(s)u_{sq0} \\ 0 & 1 - G_{PLL}(s)u_{sd0} \end{bmatrix}}^{G^v_{PLL}} \begin{bmatrix} \Delta u^s_{sd} \\ \Delta u^s_{sq} \end{bmatrix} \tag{17}$$

$$\begin{bmatrix} \Delta m^c_d \\ \Delta m^c_q \end{bmatrix} = \begin{bmatrix} \Delta m^s_d \\ \Delta m^s_q \end{bmatrix} - \overbrace{\begin{bmatrix} 0 & -G_{PLL}(s)m_{q0} \\ 0 & G_{PLL}(s)m_{d0} \end{bmatrix}}^{G^d_{PLL}} \begin{bmatrix} \Delta u^s_{sd} \\ \Delta u^s_{sq} \end{bmatrix} \tag{18}$$

where

$$tf_{PLL} = k_{pPLL} + \frac{k_{iPLL}}{s} \tag{19}$$

$$G_{PLL} = \frac{tf_{pLL}}{s + u_{sd0} f_{PLL}} \tag{20}$$

Here, tf_{PLL} is the PLL transfer function and k_{pPLL} and k_{iPLL} are the proportional and integral gains of the PLL compensator, respectively.

The linearized equation of the power balance between the AC and DC sides of the converter is expressed as

$$\Delta i_{dc} = 1.5(\begin{bmatrix} m_{d0} & m_{q10} \end{bmatrix} \begin{bmatrix} \Delta i^s_{sd} \\ \Delta i^s_{sq} \end{bmatrix} + \begin{bmatrix} i_{sd0} & i_{sq0} \end{bmatrix} \begin{bmatrix} \Delta m^s_d \\ \Delta m^s_q \end{bmatrix}) \tag{21}$$

The small signal expression of the improved virtual impedance compensator is expressed as

$$\Delta i_{dcn} = -\frac{k_i}{s} \Delta v_{dc} \tag{22}$$

$$\Delta v_{dcref} = -(R_{eq} + sL_{eq})(\frac{k_i}{s} \Delta v_{dc} + \Delta i_{dc}). \tag{23}$$

Inserting (22) and (23) into (10), the relation between the AC current reference values $\Delta i_{sd,ref}$, $\Delta i_{sq,ref}$, DC voltage Δv_{dc}, and DC current Δi_{dc} is expressed as

$$\begin{bmatrix} \Delta i_{sd,ref} \\ \Delta i_{sq,ref} \end{bmatrix} = \begin{bmatrix} H_{vdc} & H_{idc} \\ 0 & 0 \end{bmatrix} \begin{bmatrix} \Delta v_{dc} \\ \Delta i_{dc} \end{bmatrix} \tag{24}$$

where

$$H_{vdc} = -(k_{pvdc} + \frac{k_{ivdc}}{s})(\frac{k_i R_{eq}}{s} + k_i L_{eq} + 1) \tag{25}$$

$$H_{idc} = -(k_{pvdc} + \frac{k_{ivdc}}{s})(R_{eq} + sL_{eq}) \tag{26}$$

Formula (5) can be written as

$$\begin{bmatrix} \Delta i^s_{gd} \\ \Delta i^s_{gq} \end{bmatrix} = -Y_g(s) \begin{bmatrix} \Delta u^s_{sd} \\ \Delta u^s_{sq} \end{bmatrix}. \tag{27}$$

Inserting (27) into (6), the relation between the AC voltage and AC current can be given as

$$\begin{bmatrix} \Delta u^s_{sd} \\ \Delta u^s_{sq} \end{bmatrix} = \overbrace{(Y_{cf}(s) - Y_g(s))^{-1}}^{Z_s} \begin{bmatrix} \Delta i^s_{sd} \\ \Delta i^s_{sq} \end{bmatrix}. \tag{28}$$

Considering PLL, inserting (16)–(18) and (28) into (11), the modulation index can be written as

$$\begin{bmatrix} \Delta m_d^s \\ \Delta m_q^s \end{bmatrix} v_{dc0} = -G_{pwm}G_{cc}\begin{bmatrix} H_{vdc} & H_{idc} \\ 0 & 0 \end{bmatrix}\begin{bmatrix} \Delta v_{dc} \\ \Delta i_{dc} \end{bmatrix} + G_z^{vdc}(s)\begin{bmatrix} \Delta i_{sd}^s \\ \Delta i_{sq}^s \end{bmatrix} \tag{29}$$

where

$$G_z^{vdc}(s) = G_{pwm}(G_{cc} + Z_{del}) + Z_s(s)G_c^i. \tag{30}$$

$$G_c^i = (G_{pwm}G_{PLL}^v + G_{pwm}(G_{cc} + Z_{del})G_{PLL}^i + v_{dc0}G_{PLL}^d) \tag{31}$$

The relation between the DC voltage, DC current, and AC currents can be obtained from (4) by inserting (28) and (29):

$$Y_{AC}^{Vdc}(s)\begin{bmatrix} \Delta v_{dc} \\ \Delta i_{dc} \end{bmatrix} = \begin{bmatrix} \Delta i_{sd}^s \\ \Delta i_{sq}^s \end{bmatrix} \tag{32}$$

where

$$Y_{AC}^{Vdc}(s) = (Z_s(s) - Z_0(s) - G_z^{vdc}(s))^{-1}(\begin{bmatrix} m_{d0} & 0 \\ m_{q0} & 0 \end{bmatrix} + G_{pwm}G_{cc1}\begin{bmatrix} H_{vdc} & H_{idc} \\ 0 & 0 \end{bmatrix}) \tag{33}$$

and Y_{AC}^{Vdc} can be expressed as a 2*2 order matrix:

$$Y_{AC}^{Vdc} = \begin{bmatrix} Y_1 & Y_2 \\ Y_3 & Y_4 \end{bmatrix} \tag{34}$$

Inserting (32) into (4), yields

$$\begin{bmatrix} \Delta m_d^s \\ \Delta m_q^s \end{bmatrix} = \frac{1}{v_{dc0}}((Z_s(s) - Z_0(s))Y_{AC}^{Vdc}(s) - \begin{bmatrix} m_{d0} & 0 \\ m_{q0} & 0 \end{bmatrix})\begin{bmatrix} \Delta v_{dc} \\ \Delta i_{dc} \end{bmatrix}. \tag{35}$$

In the power balance relation, by inserting (34) and (35) into (21), (21) can be replaced by

$$\begin{aligned} \Delta i_{dc} &= 1.5\begin{bmatrix} m_{d0} & m_{q0} \end{bmatrix}\begin{bmatrix} \Delta i_{sd}^s \\ \Delta i_{sq}^s \end{bmatrix} + 1.5\begin{bmatrix} i_{sd0} & i_{sq0} \end{bmatrix}\begin{bmatrix} \Delta m_d^s \\ \Delta m_q^s \end{bmatrix} = 1.5\begin{bmatrix} m_{d0} & m_{q0} \end{bmatrix} Y_{AC}^{Vdc}(s)\begin{bmatrix} \Delta v_{dc} \\ \Delta i_{dc} \end{bmatrix} + \\ & 1.5\begin{bmatrix} i_{sd0} & i_{sq0} \end{bmatrix}\tfrac{1}{v_{dc0}}((Z_s(s) - Z_0(s))Y_{AC}^{Vdc}(s) - \begin{bmatrix} m_{d10} & 0 \\ m_{q10} & 0 \end{bmatrix})\begin{bmatrix} \Delta v_{dc} \\ \Delta i_{dc} \end{bmatrix} = M_1\Delta v_{dc} + M_2\Delta i_{dc} \end{aligned} \tag{36}$$

where

$$\begin{aligned} M_1 = {} & 1.5m_{d0}Y_1 + 1.5m_{q0}Y_3 + \tfrac{1.5}{v_{dc0}}i_{sd0}(RY_1 + sLY_1 - \omega_0 LY_3 - m_{d0}) + \\ & i_{sq0}(RY_3 + sLY_3 - \omega_0 LY_1 - m_{q0}) \end{aligned} \tag{37}$$

$$M_2 = 1.5m_{q0}Y_2 + \frac{1.5}{v_{dc0}}i_{sd0}(RY_2 + sLY_2 - i_{sq0}\omega_0 LY_4) + i_{sq0}(RY_2 + sLY_2 - \omega_0 LY_4) \tag{38}$$

$$R = R_n - R_c \tag{39}$$

$$L = L_n - L_c \tag{40}$$

The DC impedance of the converters can be calculated by solving (36) and can be expressed as

$$Z_{dcr} = \frac{1 - M_2}{M_1}. \tag{41}$$

Consider the DC-side capacitor,

$$Z_{dc1} = \frac{Z_{dcr}}{1 + sC_{dc}Z_{dcr}}. \tag{42}$$

3.2. DC Side Impedance Modeling of the Power-Controlled Converter

The linearized dynamic equations of the power-controlled converter are the same as those of the DC-voltage-controlled converter, which are expressed as (4)–(6), and its outer loop controller linearized equation is expressed as

$$\begin{bmatrix} \Delta i_{sd,ref} \\ \Delta i_{sq,ref} \end{bmatrix} = -H_p(\overbrace{\begin{bmatrix} 1.5u_{sd0} & 1.5u_{sq0} \\ 0 & 0 \end{bmatrix}}^{G_{vp}}\begin{bmatrix} \Delta i^c_{sd} \\ \Delta i^c_{sq} \end{bmatrix} + \overbrace{\begin{bmatrix} 1.5i_{sd0} & 1.5i_{sq0} \\ 0 & 0 \end{bmatrix}}^{G_{ip}}\begin{bmatrix} \Delta u^c_{sd} \\ \Delta u^c_{sq} \end{bmatrix}) \tag{43}$$

where $H_p = k_{pp} + k_{ip}/s$ is the active power compensator and k_{pp} and k_{ip} are the proportional and integral gains of the compensator. Its current compensator and modulator is same as (11). Inserting (16)–(18) into (42) gives

$$\begin{bmatrix} \Delta i_{sd,ref} \\ \Delta i_{sq,ref} \end{bmatrix} = -H_pG_{vp}\begin{bmatrix} \Delta i^s_{sd} \\ \Delta i^s_{sq} \end{bmatrix} - (H_pG_{vp}G^i_{PLL} + H_pG_{ip}G^i_{PLL})\begin{bmatrix} \Delta u^s_{sd} \\ \Delta u^s_{sq} \end{bmatrix}. \tag{44}$$

Inserting (16)–(18) and (28) into (11), the relation between the modulation index and AC current can be expressed as

$$\begin{aligned} \begin{bmatrix} \Delta m^s_d \\ \Delta m^s_q \end{bmatrix} v_{dc0} = {} & (G_{pwm}G_{cc}H_pG_{vp} + G_{pwm}(G_{cc} + Z_{del}))\begin{bmatrix} \Delta i^s_{sd} \\ \Delta i^s_{sq} \end{bmatrix} + \\ & (G_{pwm}G_{cc}H_p(G_{vp}G^i_{PLL} + G_{ip}G^v_{PLL}) + G^i_C)\begin{bmatrix} \Delta u^s_{sd} \\ \Delta u^s_{sq} \end{bmatrix} \end{aligned} \tag{45}$$

where

$$G^i_C = G^d_{PLL} + G_{pwm}G^v_{PLL} + G_{pwm}(G_{cc} + Z_{del})G^i_{PLL}. \tag{46}$$

Equation (45) can be rewritten as

$$\begin{bmatrix} \Delta m^s_d \\ \Delta m^s_q \end{bmatrix} v_{dc0} = G^P_Z\begin{bmatrix} \Delta i^s_{sd} \\ \Delta i^s_{sq} \end{bmatrix}. \tag{47}$$

Inserting (47) and (28) into (4) yields

$$\overbrace{(Z_s(s) - Z_0(s) - G^P_Z(s))^{-1}\begin{bmatrix} m_{d0} \\ m_{q0} \end{bmatrix}}^{Y^P_{AC}} \Delta v_{dc} = \begin{bmatrix} \Delta i^s_{sd} \\ \Delta i^s_{sq} \end{bmatrix}. \tag{48}$$

The DC-side impedance of the converter can be obtained by inserting (47) and (48) into (21):

$$Z_{dcA} = \frac{1}{1.5(\begin{bmatrix} m_{d0} & m_{q0} \end{bmatrix} + \begin{bmatrix} i_{sd0} & i_{sq0} \end{bmatrix} G^P_Z)Y^P_{AC}}. \tag{49}$$

Consider the DC-side capacitor,

$$Z_{dcB} = \frac{Z_{dcA}}{1 + sC_{dc}Z_{dcA}} \tag{50}$$

Adding the DC cable impedance to the DC-side impedance of the power-controlled converter yields

$$Z_{dc2}(s) = (Z_{dcB}(s) \Big\| \frac{2}{sC_d} + sL_d + R_d) \Big\| \frac{2}{sC_d}. \tag{51}$$

3.3. Verifying Impedance Modeling Through Perturbation Signal Testing

Perturbation signal testing is used to verify the accuracy of the proposed small-signal model. An AC current source should be placed parallel to the DC side of the system as an input signal, as Figure 6a shows. It is necessary to inject AC current at different frequencies to measure the DC side impedance at different frequencies. At each injection frequency, a simulation experiment is conducted. The DC voltage and DC current data of each experiment are analyzed by fast Fourier transformation. The components under the disturbance frequency are taken out, and the ratio of DC voltage and DC current is calculated as the calculated impedance [13]. Figure 6b,c shows the DC-side impedance verification of the DC system rectifier and inverter sides, respectively. The solid line in Figure 6b represents the analytical impedance of the rectifier converter as in (42), and the points represent the simulation results. Figure 6b represents the analytical impedance of the inverter station and the DC cable as (51), and the points represent the simulation results.

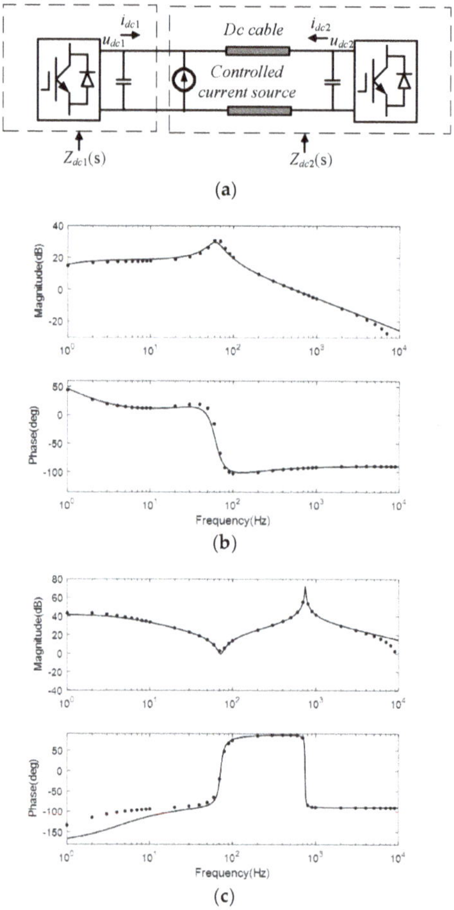

Figure 6. Frequency response of the impedance and verification. The solid line represents the model prediction, and the black points denote the simulation. (**a**) Disturbance signal testing. (**b**) Impedance of the DC voltage-controlled converter (**c**). Impedance of the power-controlled converter.

4. Stability Analysis of VSC–HVDC with Improved Virtual Impedance Control Strategy

According to the results of small signal modeling, the DC side of the voltage-controlled converter is modeled by a DC voltage source (V_s), in series with an output impedance (Z_s), which equals to Z_{dc1}. The DC side of the power-controlled converter and DC cable is modeled by a DC current source shunted with an input impedance (Z_l), which equals to Z_{dc2}. Figure 7 shows the equivalent impedance model of the system.

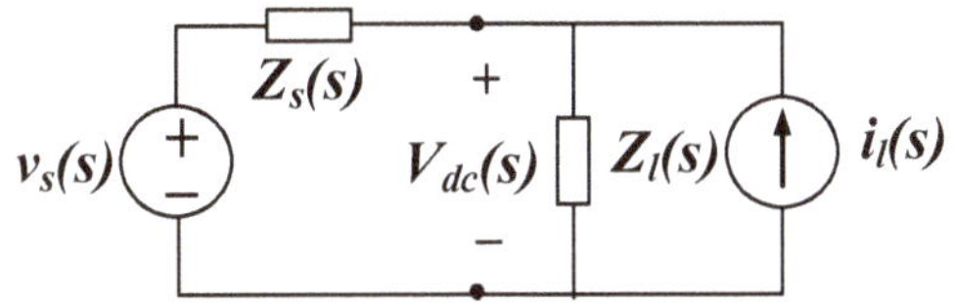

Figure 7. Equivalent model of the VSC–HVDC system.

According to Kirchhoff's law, the output voltage of the DC side V_{dc} (s) can be expressed by Formula (52). The stability of the DC system depends on the ratio of Z_s to Z_l, which is the open-loop transfer function of system T_m. DC voltage is predicted to be stable when T_m satisfies the Nyquist stability criterion [12–14]:

$$V_{dc}(s) = (v_s(s) + i_l(s)Z_s(s))(\frac{1}{1+T_m}) \tag{52}$$

$$T_m = \frac{Z_s(s)}{Z_l(s)}. \tag{53}$$

4.1. *Impact of the Power Flow Direction*

Figure 8a shows the Nichols plots of an open-loop transfer function T_m when the transmission power is ±500 MW and the length of the DC cable is 50 km. The traditional control strategy [13] shown in Figure 3 is adopted to a DC-voltage-controlled converter, and controller parameters are set as $k_{pvdc} = 15$, $k_{ivdc} = 100$.

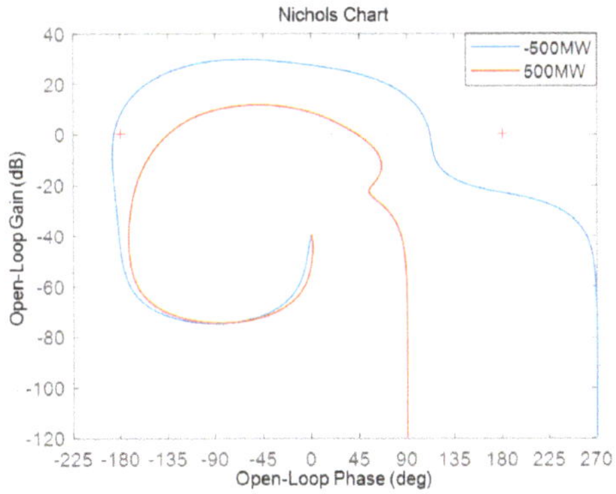

(a)

Figure 8. *Cont.*

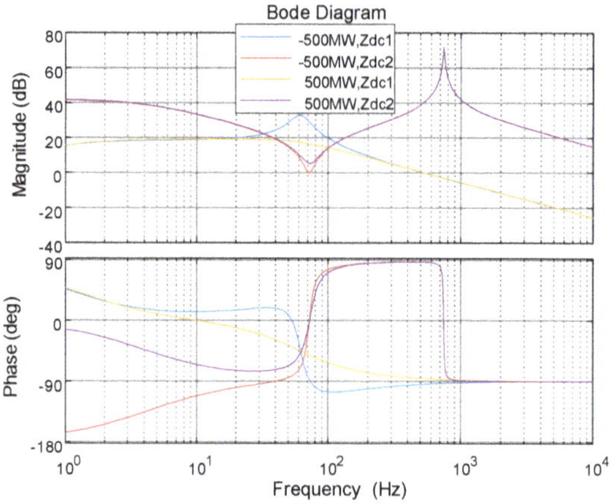

(b)

Figure 8. (**a**) Nichols plots of T_m when transmission power is ±500 MW. (**b**) Impedance frequency responses of Z_{dc1} and Z_{dc2} when transmission power is ±500 MW.

As shown in Figure 8a, the red line does not encircle (−180°, 0), and the VSC–HVDC system is predicted to be stable. The blue encircles (−180°, 0), and the VSC–HVDC system is predicted to be unstable.

Figure 8b shows the impedance frequency responses of Z_{dc1} and Z_{dc2} when the active power is set as ±500 MW. Compared with the yellow line and the blue line, in the mid-frequency band, the blue line shows negative damping while the yellow line does not. Z_{dc1} and Z_{dc2} intersect at the mid-frequency band due to the influence of DC cable impedance. The phase difference at the red line and blue line in the mid-frequency band is approximately 180°, so the DC voltage is predicted to oscillate. The stability analysis results are consistent with the simulation results in Figure 4.

4.2. Impact of Virtual Impedance by Frequency Responses

Figure 9 shows the impedance frequency responses of Z_{dc1} and Z_{dc2} when the active power is set as −500 MW and the length of DC cable is 50 km. The proposed strategy in this paper is adopted for a DC-voltage-controlled converter, and the controller parameters are set as k_{pvdc} = 15, k_{ivdc} = 100, and k_i = 10. Figure 9a shows the impedance frequency responses of Z_{dc1} under a different R_{eq}s. The pink line represents Z_{dc2} and the blue, red, and yellow lines represents Z_{dc1} when R_{eq} = 0.5, 2 and 5, respectively. The blue line shows that when the phase of Z_{dc1} impedance in the mid-frequency band is below −90°, the system exhibits negative damping. The phase difference at the intersection of the pink and blue lines in the mid-frequency band is approximately 180°, and the DC voltage is predicted to oscillate. The red and yellow lines show that when the phase of Z_{dc1} impedance in the mid-frequency band is above -90°, and the phase difference is less than 160 degrees, the DC voltage is predicted to stable. However, with the increase of R_{eq}, the phase difference at the intersection of the pink and blue lines in low-frequency bands increases, and the system tends to lose stability.

Figure 9b shows impedance frequency responses of Z_{dc1} under a different L_{eq}s. The pink line represents Z_{dc2} and the blue, red, and yellow lines represent Z_{dc1} when L_{eq} = 0.002, 0.02, and 0.05, respectively. Blue line shows that phase of Z_{dc1} impedance in mid-frequency band is below −90°, the system exhibits negative damping. The phase difference at the intersection of the pink and blue lines in the mid-frequency band is approximately 180°, and the DC voltage is predicted to oscillate. Red and yellow lines show that when the phase of Z_{dc1} impedance in the mid-frequency band is above −90°, and the phase difference is less than 160 degrees, the DC voltage is predicted to be stable.

However, with the increase of L_{eq}, the phase difference at the intersection of the pink and blue lines in the low-frequency band increases, and the system tends to lose stability.

The conclusion is that if the virtual impedance is too small, it will not be enough to suppress the oscillation. If the virtual impedance is too large, a new oscillation may occur in the low frequency band.

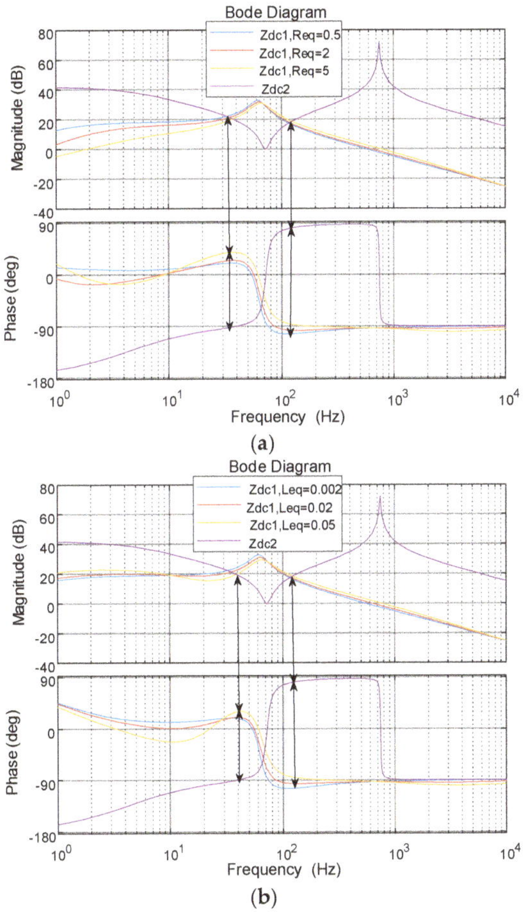

Figure 9. Impedance frequency responses of Z_{dc1} and Z_{dc2} under different equivalent virtual impedance values (**a**) R_{eq} and (**b**) L_{eq}.

4.3. Impact of Virtual Impedance Parameters by Nichols Plots

Figure 10 shows the Nichols plots of T_{m} with different virtual impedance parameters. The DC cable length is 10 km and the transmission power is −500 MW. The proposed strategy in this paper is adopted to the DC-voltage-controlled converter, and the controller parameters are set as k_{pvdc} = 15, k_{ivdc} = 100, and k_i = 10. The blue, red, and yellow lines in Figure 10a show the Nichols plots of T_{m} when the virtual impedance parameters R_{eq} are 0.5 Ω, 2 Ω, and 5 Ω, respectively. Figure 10a shows that T_{m} encircles (−180°, 0), and the system is predicted to be unstable when R_{eq} = 0.5 Ω. The system is predicted to be stable when R_{eq} = 2 and 5 Ω. Increasing the R_{eq} value in a certain range helps improve the stability of the system. The phase margin of the system is insufficient to suppress DC-side oscillation when R_{eq} = 0.5 Ω.

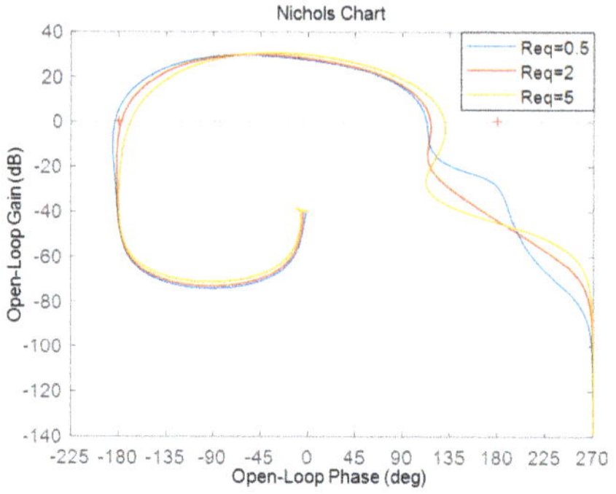

(a)

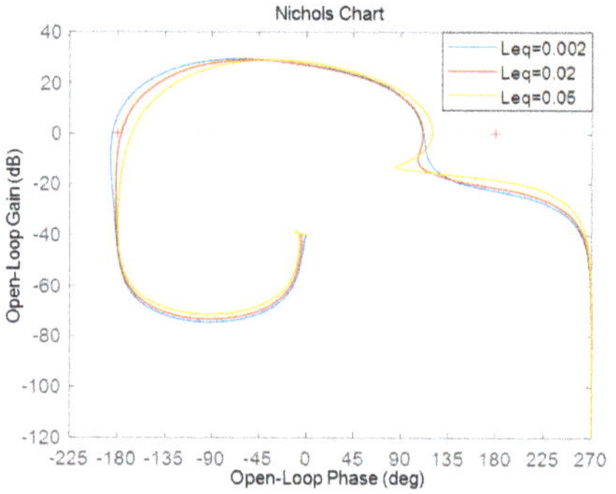

(b)

Figure 10. Nichols plots of T_m under different equivalent virtual impedance values (**a**) R_{eq}; (**b**) L_{eq}.

The blue, red, and yellow lines in Figure 10b show the Nichols plots of T_{m} when virtual impedance parameters L_{eq} are 0.002 Ω, 0.02 Ω, and 0.05 Ω, respectively. Figure 10b shows that T_{m} encircles (−180°, 0), and the system is predicted to be unstable when L_{eq} = 0.002 Ω. The system is predicted to be stable when L_{eq} = 0.02 and 0.05 Ω. Increasing the L_{eq} value in a certain range helps improve the stability of the system. The phase margin of the system is insufficient to suppress DC-side oscillation when L_{eq} = 0.002 Ω.

4.4. Impact of DC Cable Length

Figure 11 shows the Nichols plots of the open-loop transfer function T_m under different lengths of DC cable when the transmission power is −500 MW. The control parameters are k_{pvdc} = 15, k_{ivdc} = 100, k_i = 10, R_{eq} = 0.5 Ω, and L_{eq} = 0.002 Ω. The blue, red and yellow lines in Figure 11 represent the Nichols plots of T_m at DC cable lengths of 50, 100, and 150 km, respectively. As shown in Figure 11, the Nichols plot gradually approaches (−180°, 0), with a decrease in DC cable length. The Nichols plot encircles (−180°, 0), and the system is predicted to be unstable when the DC cable length is reduced to 50 km.

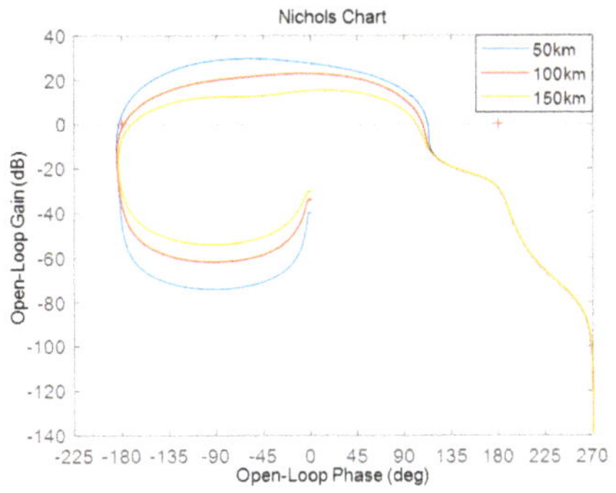

Figure 11. Nichols plots of T_m under different DC cable lengths.

4.5. Impact of DC Side Capacity

Figure 12 shows the Nichols plots of the open-loop transfer function T_m under a different capacitance of the DC side capacity when the transmission power is −500 MW. The control parameters are k_{pvdc} = 15, k_{ivdc} = 100, k_i = 10, R_{eq} = 0.5 Ω, and L_{eq} = 0.002 Ω. The blue, red, and yellow lines in Figure 12 represent the Nichols plots of T_m at the DC side capacity of 300, 450, and 600 μf, respectively. As shown in Figure 12, the Nichols plot gradually approaches (−180°, 0), with a decrease in the DC side capacity. The Nichols plot encircles (−180°, 0), and the system is predicted to be unstable when the DC side capacity is reduced to 300 μf.

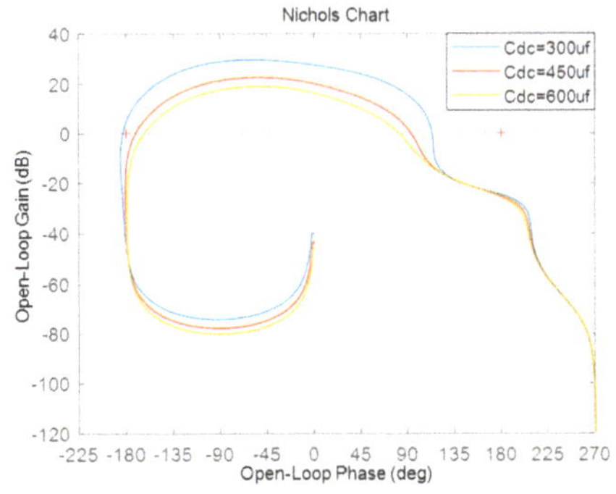

Figure 12. Nichols plots of T_m under different DC side capacities.

From the above stability analysis, it can be concluded that there are three main factors affecting the stability of the system—power flow [24], main circuit parameters, and controller parameters. Among the main circuit parameters, the most important factors are the length of the circuit and the capacitance of the DC side. According to the stability, analysis results in 4.4 and 4.5. The stability margin of the system can be improved by increasing the DC capacitance and DC cable length. When the main circuit parameters are determined and the power flow is determined, the stability of the system can also be improved by optimizing the controller settings.

4.6. Impact of Grid Impedance

Figure 13 shows the Nichols plots of the open-loop transfer function T_m under different grid impedance when the transmission power is −500 MW. The control parameters are k_{pvdc} = 15, k_{ivdc} = 100, k_i = 10, and L_{eq} = 0.02 Ω. The blue and red lines in Figure 13 represent the Nichols plots of T_m when L_n = 1 and 5 mH, respectively. The blue line does not encircle (−180°, 0), while the red line encircles

(−180°, 0), which means and DC-side oscillations are more likely to occur in weaker power grids. The yellow line in Figure 13 represents the Nichols plot of T_m when L_n = 5 mH and R_{eq} = 2. The yellow line does not encircle (−180°, 0), which means the system is predicted to be stable after choosing appropriate virtual impedance parameters.

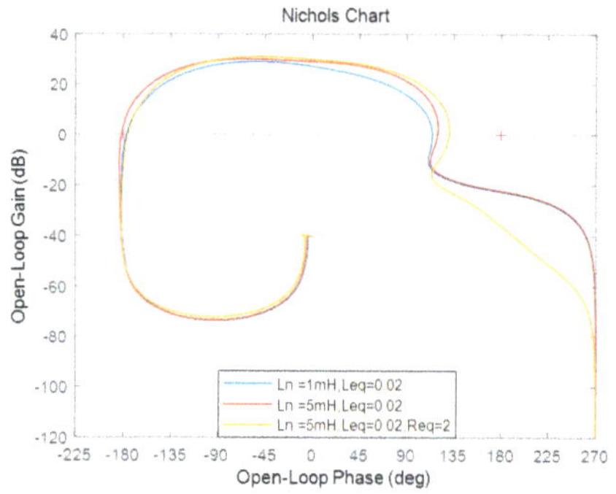

Figure 13. Nichols plots of T_m under a different grid impedance.

5. Simulation Verification

5.1. Impact of DC Cable Length

Figure 14 shows the simulation results of the VSC–HVDC system with virtual impedance control strategy under different DC cable lengths. The voltage control parameters are k_{pvdc} = 15, k_{ivdc} = 100, k_i = 10, R_{eq} = 0.5 Ω, and L_{eq} = 0.002 Ω. At 2 s, the steady-state power command value is set from 500 MW to −500 MW. As shown in Figure 14, the blue line indicates the simulation results when the DC cable length is 50 km, and the red line indicates 100 km. The DC voltage in the blue line drops by approximately 5% and begins to oscillate, and the system loses stability. The DC voltage in the red line drops by approximately 5%, smoothly restoring the instruction value, and the system remains stable. The simulation results are consistent with the theoretical analysis in Section 4.4.

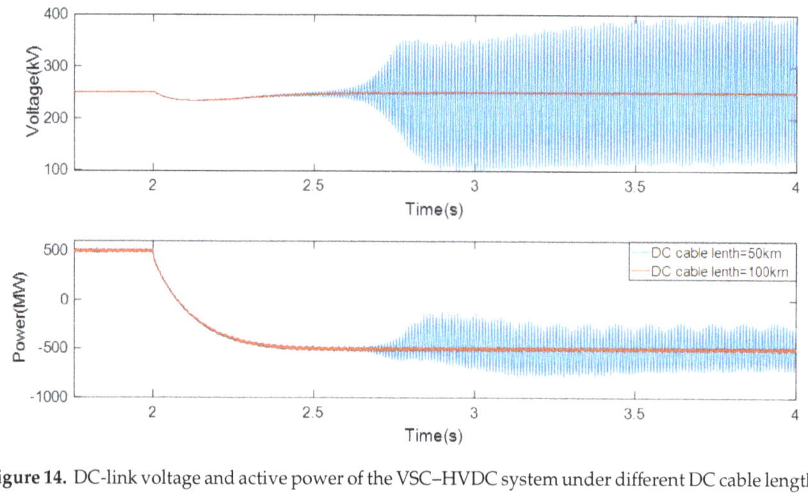

Figure 14. DC-link voltage and active power of the VSC–HVDC system under different DC cable lengths.

5.2. Impact of DC Side Capacity

Figure 15 shows the simulation results of the VSC–HVDC system with a virtual impedance control strategy under a different DC side capacity. The voltage control parameters are k_{pvdc} = 15, k_{ivdc} = 100, k_i = 10, R_{eq} = 0.5 Ω, and L_{eq} = 0.002 Ω. At 2 s, the steady-state power command value is set from 500 MW to −500 MW. As shown in Figure 15, the blue line indicates the simulation results when the DC side capacity is 300 µf, and the red line indicates 450 µf. The DC voltage in the blue line drops by approximately 5% and begins to oscillate, and the system loses stability. The DC voltage in the red line drops by approximately 5%, smoothly restoring the instruction value, and the system remains stable. The simulation results are consistent with the theoretical analysis in Section 4.5.

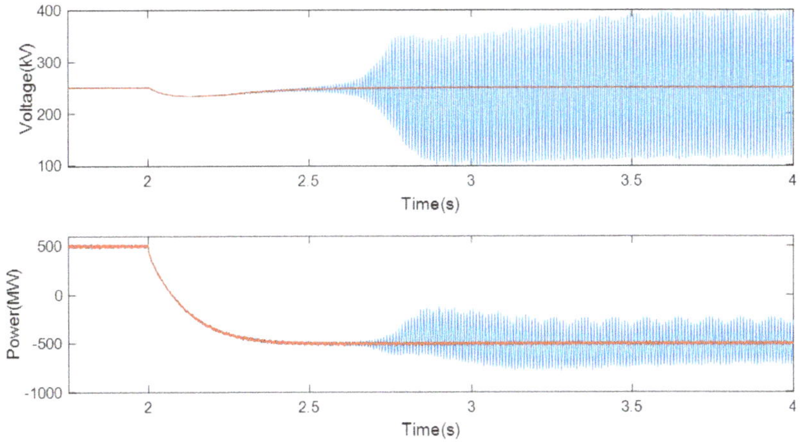

Figure 15. DC-link voltage and active power of the VSC–HVDC system under different DC side capacity.

5.3. Dynamic Performance Comparison

According to [13], reducing the value of k_{pvdc} can improve the output impedance of the DC-voltage-controlled converter, thereby reinforcing the stability of the system. However, this condition influences the dynamic performance of the control. This paper compares the dynamic performance of the two control strategies. Similarly, the DC cable length is 50 km, and the active power instruction changes from 500 MW to −500 MW at 2 s. As shown in Figure 16a, the blue line represents the DC voltage under the traditional control strategy [13]. When k_{pvdc} = 5 and k_{ivdc} = 100, the red line represents the DC voltage under the proposed control strategy when k_{pvdc} = 15, k_{ivdc} = 100, k_i = 10 and R_{eq} = 2 Ω, and the yellow line represents the DC voltage under the proposed control strategy when k_{pvdc} = 15, k_{ivdc} = 10, k_i = 10, and L_{eq} = 0.02 Ω. The DC voltage in the blue line drops by approximately 9% when the instruction value is changed and gradually increases to the instruction value. The DC voltage in the red line drops by only 2% when the instruction value is changed and increases the instruction value by more than 2%. The DC voltage in the yellow line drops by only 4% when the instruction value is changed and gradually increases to the instruction value. The dynamic performance of the proposed control strategy is better than that of the traditional control method.

Figure 16b compares system DC voltages under the proposed control strategy with different virtual impedance parameters when k_{pvdc} = 15, k_{ivdc} = 100 and k_i = 10. The blue line represents R_{eq} = 0.5, L_{eq} = 0.02 Ω, the red line represents R_{eq} = 2 Ω, L_{eq} = 0.02 Ω, and the yellow line represents R_{eq} = 2 Ω, L_{eq} = 0.002 Ω. From the comparison of these three lines, it can be seen that the speed of DC voltage regulation mainly depends on R_{eq}. As can be seen from Figure 9a, increasing R_{eq} can reduce the amplitude response of Z_{dc1} in the low-frequency band. However, increasing L_{eq} has little effect on the amplitude response of Z_{dc1} in the low-frequency band, as in Figure 9b.

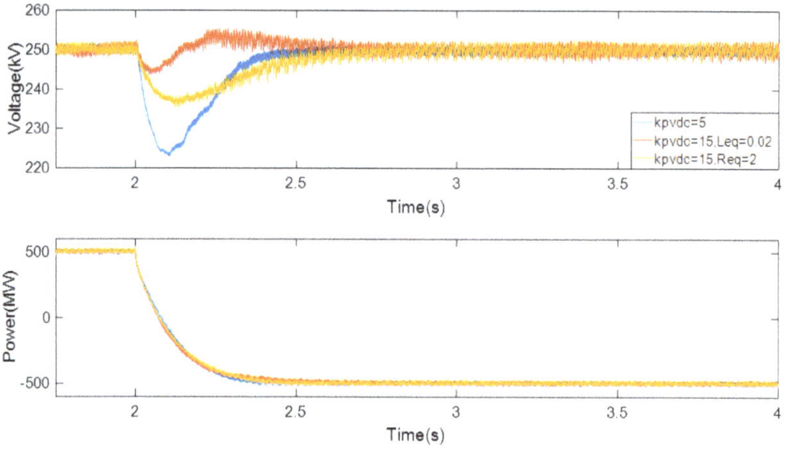

(a)

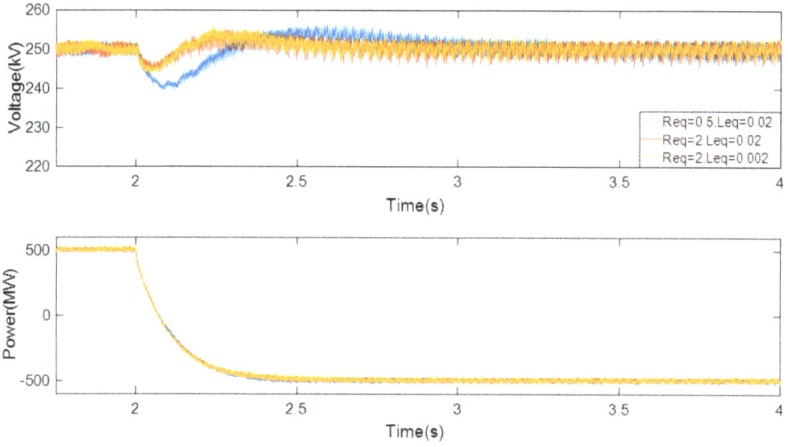

(b)

Figure 16. DC-link voltage and active power of the VSC–HVDC system (**a**) under the proposed control strategy and traditional control strategy [13] and (**b**) under the proposed control strategy with different virtual impedance parameters.

5.4. Steady State Error Elimination

On the basis of the previous analysis, although the traditional virtual impedance control strategy can enhance the transmission capacity of the system, it causes the steady-state error of DC voltage. Figure 17 shows the comparison of two virtual impedance control strategies. The length of the DC cable is 50 km. At 2 s, the steady-state power command value is set from 500 MW to −500 MW.

The DC voltage in the blue line represents the traditional virtual impedance, shown as the yellow frame in Figure 5, and the controller parameters are k_{pvdc} = 15, k_{ivdc} = 100, and R_{eq} = 2 Ω. The DC voltage in the red line represents the proposed control strategy, with k_{pvdc} = 15, k_{ivdc} = 100, R_{eq} = 2 Ω, and k_i = 10. Under the two control strategies, the DC voltage fluctuates in a short time and a small range only, and the active power can steadily step down. The DC voltage shown by the red line is close to the instruction value of 250 kV, and the steady-state error of the blue line is approximately 5 kV in both power flow directions, which are consistent with the theoretical analysis. The simulation results show that the improved virtual impedance control method can improve the stability of the DC system and eliminate the steady-state error of the DC voltage caused by the virtual impedance.

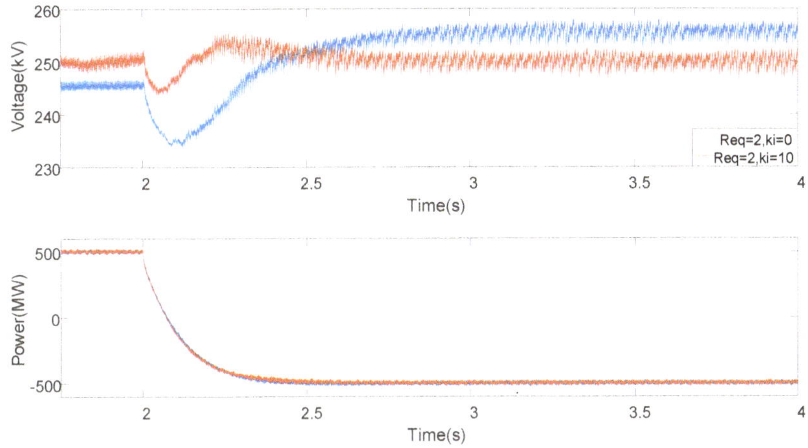

Figure 17. DC-link voltage and active power of the VSC–HVDC system under the proposed control strategy and traditional virtual impedance.

5.5. *Impact of Gird Impedance*

Figure 18 shows the simulation results of the VSC–HVDC system with a virtual impedance control strategy under a different grid impedance. The voltage control parameters are k_{pvdc} = 15, k_{ivdc} = 100, k_i = 10, and the transmission power is set as −500 MW. At 2 s, the grid impedance is switched from 1 mH to 5 mH. The blue line in Figure 18 indicates the simulation results when the virtual impedance parameter is L_{eq} = 0.02 Ω. The DC voltage in the blue line begins to oscillate, and the system loses stability. The red line in Figure 18 indicates the simulation results when virtual impedance parameters are L_{eq} = 0.02 Ω and R_{eq} = 2 Ω. The DC voltage in the red line stays stable, and the system remains stable. The simulation results are consistent with the theoretical analysis in Section 4.6.

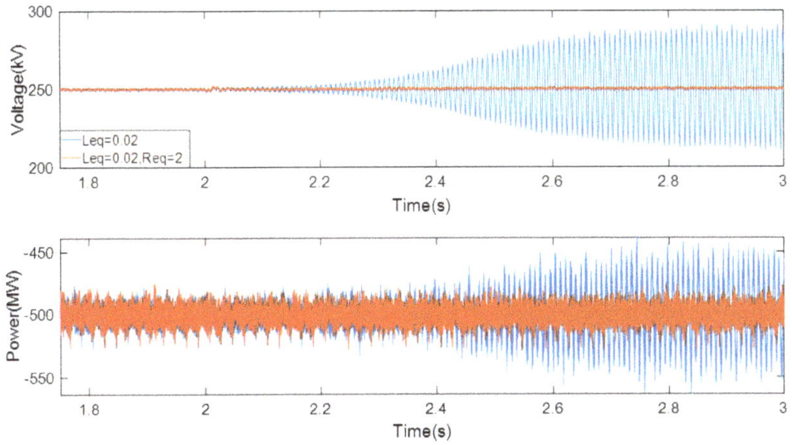

Figure 18. DC-link voltage and active power of the VSC–HVDC system under different gird impedance.

6. Conclusions

This paper investigates the stability of VSC–HVDC operation in a bidirectional power flow mode. DC cable length affects the reverse power transmission power of the system. An improved virtual impedance control strategy is presented to increase the reverse power transmission power of the system. A system stability analysis based on the impedance model is applied to the proposed control strategy. System stability is affected by the power flow, controller parameters, DC cable length, and DC side capacity.

(1) DC-side oscillation occurs when the transmission power of the system is large. The maximum transmission power of a DC voltage-controlled converter to a power-controlled converter is less than that in the opposite transmission direction.
(2) The shorter the DC cable is, the more easily the oscillation of DC voltage will occur.
(3) The smaller the DC side capacity is, the more easily the oscillation of DC voltage will occur.
(4) The weaker the AC grid strength is, the more easily the DC side oscillation will occur.
(5) Appropriate virtual impedance parameters can improve system stability. The phase margin of the system is insufficient to suppress DC-side oscillation when the virtual impedance parameter is small. If the virtual impedance parameters, R_{eq} or L_{eq}, are too large, then the system will enter a new unstable state.

The simulation results show that the proposed improved control strategy can eliminate the steady-state error of DC voltage caused by virtual impedance and maintain the advantage of a virtual impedance strategy to improve system stability. The proposed control method has better dynamic performance compared to the traditional control method.

Author Contributions: All the authors have contributed to this paper in different aspects. Y.L. proposed the original concept and wrote the original draft. K.L. acted as supervisor and gave suggestions on paper improvement. S.Z., X.L., and Q.H. helped with software simulation and methodology improvement.

Funding: This research was funded by [Nation Key Research and Development Program of China] grant number [2017YFB0903700] And The APC was funded by [2017YFB0903700].

Conflicts of Interest: The authors declare no conflict of interest.

References

1. Flourentzou, N.; Agelidis, V.G.; Demetriades, G.D. VSC-Based HVDC Power Transmission Systems: An Overview. *IEEE Trans. Power Electron.* **2009**, *24*, 592–602. [CrossRef]
2. Yang, J.; Liu, K.; Yu, Y.; Qin, L.; Le, J. Small Signal Modeling for VSC-HVDC Used in AC Grid Interconnection. *Proc. CSEE* **2015**, *35*, 2177–2184.
3. Zhang, L.; Nee, H.P. Multivariable feedback design of VSC-HVDC connected to weak ac systems. In Proceedings of the IEEE Bucharest PowerTech, Bucharest, Romania, 28 June–2 July 2009; pp. 1–8.
4. Kumars, R.; Arash, M.; Jose, I.C.; Alvaro, L.; Pedro, R. A Generalized Voltage Droop Strategy for Control of Multiterminal DC Grids. *IEEE Trans. Ind. Appl.* **2015**, *51*, 607–618.
5. Zhang, L.; Nee, H.P.; Harnefors, L. Analysis of Stability Limitations of a VSC-HVDC Link Using Power-Synchronization Control. *IEEE Trans. Power Syst.* **2011**, *26*, 1326–1337. [CrossRef]
6. Zhang, L.; Harnefors, L.; Nee, H.P. Interconnection of Two Very Weak AC Systems by VSC-HVDC Links Using Power-Synchronization Control. *IEEE Trans. Power Syst.* **2011**, *26*, 344–355. [CrossRef]
7. Wen, B.; Dong, D.; Boroyevich, D.; Mattavelli, P.; Shen, Z. Impedance-Based Analysis of Grid-Synchronization Stability for Three-Phase Paralleled Converters. *IEEE Trans. Power Electron.* **2015**, *31*, 675–687. [CrossRef]
8. Yuan, H.; Yuan, X.; Hu, J. Modeling of Grid-Connected VSCs for Power System Small-Signal Stability Analysis in DC-Link Voltage Control Time-scale. *IEEE Trans. Power Syst.* **2017**, *32*, 3981–3991. [CrossRef]
9. Kalcon, G.O.; Adam, G.P.; Anaya-Lara, O.; Lo, S.; Uhlen, K. Small-Signal Stability Analysis of Multi-Terminal VSC-Based DC Transmission Systems. *IEEE Trans. Power Syst.* **2012**, *27*, 1818–1830. [CrossRef]
10. Beerten, J.; D'Arco, S.; Suul, J.A. Identification and Small-Signal Analysis of Interaction Modes in VSC MTDC Systems. *IEEE Trans. Power Deliv.* **2016**, *31*, 888–897. [CrossRef]

11. Li, Y.; Tang, G.; He, Z.; An, T.; Yang, J.; Wu, Y.; Kong, M. Damping Control Strategy research for MMC based HVDC system. *Proc. CSEE* **2016**, *36*, 5492–5503.
12. Xu, L.; Fan, L.; Miao, Z. DC Impedance-Model-Based Resonance Analysis of a VSC–HVDC System. *IEEE Trans. Power Deliv.* **2014**, *30*, 1221–1230. [CrossRef]
13. Amin, M.; Molinas, M.; Lyu, J.; Cai, X. Impact of Power Flow Direction on the Stability of VSC-HVDC Seen from the Impedances Nyquist Plot. *IEEE Trans. Power Electron.* **2016**, *32*, 8204–8217. [CrossRef]
14. Middlebrook, R.D. Input Filter considerations In Design and Application of Switching Regulators. In Proceedings of the IEEE Power Electronics Specialists Conference, Cleveland, OH, USA, 8 June 1976; pp. 366–382.
15. Sun, J. Impedance-Based Stability Criterion for Grid-Connected Inverters. *IEEE Trans. Power Electron.* **2011**, *26*, 3075–3078. [CrossRef]
16. Pinares, G.; Bongiorno, M. Modeling and Analysis of VSC-Based HVDC Systems for DC Network Stability Studies. *IEEE Trans. Power Deliv.* **2015**, *31*, 848–856. [CrossRef]
17. Pinares, G.; Bongiorno, M. Analysis and Mitigation of Instabilities Originated From DC-Side Resonances in VSC-HVDC Systems. *IEEE Trans. Ind. Appl.* **2016**, *52*, 2807–2815. [CrossRef]
18. Song, Y.; Breitholtz, C. Nyquist Stability Analysis of an AC-Grid Connected VSC-HVDC System Using a Distributed Parameter DC Cable Model. *IEEE Trans. Power Deliv.* **2016**, *31*, 898–907. [CrossRef]
19. Amin, M.; Molinas, M. Small-Signal Stability Assessment of Power Electronics Based Power Systems: A Discussion of Impedance- and Eigenvalue-Based Methods. *IEEE Trans. Ind. Appl.* **2017**, *53*, 5014–5030. [CrossRef]
20. Yu, X.; Salato, M. An Optimal Minimum-Component DC–DC Converter Input Filter Design and Its Stability Analysis. *IEEE Trans. Power Electron.* **2014**, *29*, 829–840.
21. Cespedes, M.; Xing, L.; Sun, J. Constant-Power Load System Stabilization by Passive Damping. *IEEE Trans. Power Electron.* **2011**, *26*, 1832–1836. [CrossRef]
22. Radwan, A.A.A.; Mohamed, A.R.I. Linear Active Stabilization of Converter-Dominated DC Microgrids. *IEEE Trans. Smart Grid.* **2012**, *3*, 203–216. [CrossRef]
23. Wu, M.; Lu, D.C. A Novel Stabilization Method of LC Input Filter With Constant Power Loads Without Load Performance Compromise in DC Microgrids. *IEEE Trans. Ind. Electron.* **2015**, *62*, 4552–4562. [CrossRef]
24. Wu, W.; Chen, Y.; Luo, A.; Zhou, L.; Zhou, X.; Yang, L.; Huang, X. A Virtual Phase-Lead Impedance Stability Control Strategy for the Maritime VSC-HVDC System. *IEEE Trans. Ind. Inf.* **2018**, *14*, 5475–5486. [CrossRef]
25. Tang, X.; Zhang, K.; Xu, Q.; Chen, S.; Tan, W. Control strategy for enlarging the transmission capacity of VSC-HVDC systems supplying passive networks. *Trans. China Electron.* **2016**, *31*, 44–51.
26. Paquette, A.D.; Divan, D.M. Virtual Impedance Current Limiting for Inverters in Microgrids With Synchronous Generators. *IEEE Trans. Ind. Appl.* **2015**, *51*, 1630–1638. [CrossRef]
27. Lu, X.; Wang, J.; Guerrero, J.M.; Zhao, D. Virtual-Impedance-Based Fault Current Limiters for Inverter Dominated AC Microgrids. *IEEE Trans. Smart Grid* **2018**, *9*, 1599–1612. [CrossRef]
28. Rizqiawan, A.; Fujita, G.; Funabashi, T.; Nomura, M. Impact of a virtual impedance on the input admittance of a grid—Connected inverter. *IEEE Trans. Electr. Electron.* **2013**, *8*, 190–198. [CrossRef]
29. Li, Y.; Tang, G.; Ge, J.; He, Z.; Pang, H.; Yang, J.; Wu, Y. Modeling and Damping Control of Modular Multilevel Converter Based DC Grid. *IEEE Trans. Power Syst.* **2018**, *33*, 723–735. [CrossRef]

applied sciences

MDPI

Article

Novel Transient Power Control Schemes for BTB VSCs to Improve Angle Stability

Sungyoon Song [1], Sungchul Hwang [1], Baekkyeong Ko [2], Seungtae Cha [2] and Gilsoo Jang [1,*]

1 School of Electrical Engineering, Korea University, Anam-ro, Sungbuk-gu, Seoul 02841, Korea; blue6947@korea.ac.kr (S.S.); adidas@korea.ac.kr (S.H.)

2 Korea Electric Power Corporation (KEPCO), Naju-si, Jeollanam-do, 58322, Korea; bk.ko@kepco.co.kr (B.K.); seungtae.cha@kepco.co.kr (S.C.)

* Correspondence: gjang@korea.ac.kr; Tel.: +82-010-3412-2605

Received: 20 June 2018; Accepted: 8 August 2018; Published: 11 August 2018

Abstract: This paper proposes two novel power control strategies to improve the angle stability of generators using a Back-to-Back (BTB) system-based voltage source converter (VSC). The proposed power control strategies have two communication systems: a bus angle monitoring system and a special protection system (SPS), respectively. The first power control strategy can emulate the behaviour of the ac transmission to improve the angle stability while supporting the ac voltage at the primary level of the control structure. The second power control scheme uses an SPS signal to contribute stability to the power system under severe contingencies involving the other generators. The results for the proposed control scheme were validated using the PSS/E software package with a sub-module written in the Python language, and the simple assistant power control with two communication systems is shown to improve the angle stability. In conclusion, BTB VSCs can contribute their power control strategies to ac grid in addition to offering several existing advantages, which makes them applicable for use in the commensurate protection of large ac grid.

Keywords: BTB-HVDC; power control; angle stability; special protection system

1. Introduction

Back-to-Back system-based voltage source converters (BTB VSC) have been developed in numerous studies and are increasingly installed in ac grids [1,2]. The system is generally installed to improve the connection point of a renewable energy system, since the converter plays an important role in voltage swings during a contingency while decreasing the magnitude of the fault current in the ac grid. Another advantage of an embedded BTB system is that it increases the ac transmission transfer capacity by supporting the voltage and reduces the angle differences between each side of ac system. In the transient period, the system also contributes to voltage stability by applying ac voltage control strategy while sustaining their transfer power [3]. The ac voltage and the active power control are essential operational tasks in BTB VSCs to guarantee the proper stability of ac grid. However, the BTB VSCs generally sustain their fixed active power even though the ac grid has severe contingencies due to the fact that, unlike a Point-to-Point (PTP) system, frequency control with an embedded BTB system is useless in a large ac grid. Thus, novel transient power control schemes including two communication systems are newly introduced in this paper to expand the advantage of BTB VSCs. The proposed schemes contribute angle stability while also providing reactive power for voltage stability.

The BTB VSCs can further contribute transient stability using their advantage of flexible control. In fact, in the previous study, two kinds of transient power control methods were identified in the CIGRE (International Council on Large Electric Systems) conference, and one of them involves the use of WAMS (Wide Area Measurement Systems) by measuring the voltage and current from a GPS (Global Positioning System) system to prevent damping oscillation or an overload line problem [4].

The second method allows the power to increase its desired set point or to be automatically reduced as parallel ac transmission in the post-disturbance period. In detail, the active power reference of the BTB VSC is modified by the same amount as the parallel ac transmission line after the fault to resolve an overload situation. However, the first method has a major problem: the installation cost of entire grid monitoring system is too high. The second method lacks improvements in the transient response, and this type of control may not be practical given that the absence of the parallel ac transmission line restricts the use of power control. Furthermore, the overload problem is not eliminated at all positions according to the BTB system installation point.

The previous studies related to transient power control with BTB VSC has not been specifically illustrated, since the conservative operation with power control is general [3]. In the future, however, as the volatility of renewable energy sources increases, these Flexible AC transmission system (FACTS) will be required to be more flexible. To maximize the strengths of the BTB VSC, different form of control strategy is needed. Note that the ac voltage control is an ideal way to prevent the voltage instability in the embedded BTB VSC system, only power control is discussed in this paper. The first power control method using bus angle information acts like a frequency control in PTP HVDC, and it contributes the angle stability of several generators. It also finds a new convergence point immediately after a contingency such as ac transmission lines. The second power control strategy using a special protection system (SPS) signal could cause the tripped generator be reduced during the transient period, and it makes grid operation more flexible. The most important contribution of the proposed methods is that it can alleviate the first damping of the generators' angle and prevent the rest from a potential loss of synchronism. To perform the analysis, in Section 2, the basic model of the VSCs is derived. In Section 3, two transient power control models are configured. Lastly, the angle stability evaluation using the proposed power control schemes with BTB VSCs is represented in Section 4. The angle spread analysis regarding first damping in PSS/E software is used to demonstrate the effectiveness of the proposed model.

2. VSC Model Configuration

2.1. VSC Configuration

Based on the Insulated Gate Bipolar Transistor (IGBT) and Pulse-Wide Modulation (PWM) skill, the VSC is capable of yielding a high active and reactive power input to the grid independently in a low grid voltage situation. Nowadays, Modular Multilevel Converter (MMC), using cascaded connection logic, is more attractive for application in the grid because of its unique features, e.g., the good sinusoidal waveform of its output voltage and low switching loss. Manufacturers have developed new generation of VSC based on MMC, and is has fast response speed, especially during the transient phase after disturbances such as voltage swing. In this paper, the MMC based BTB VSC system will be discussed.

In the BTB VSC system, the two converter stations are located at the same site, and the two ac systems are interconnected with either the same or a different frequency. The control of both active and reactive power is bi-directional and across the entire capability range. Within the capability curve, the VSC located in a weak ac system can support frequency and voltage drops and responses to several disturbances with fast dynamic control. To apply the bi-directional control of the converter, the ac current and voltage have to be transformed into a rotating direct-quadrature (d-q) frame. The rotating reference frame is aligned to the voltage phasor of the point of common coupling (PCC). Since the q-axis term of the voltage approaches zero, the converter enables the decoupled control of active and reactive power [5].

As we can be observed in Figure 1, the PCC voltage (e_c) and converter voltage (v_c) can be expressed with the resistance and inductance.

$$e_c - v_c = R_T i_c + L_T \frac{di_c}{dt}, \tag{1}$$

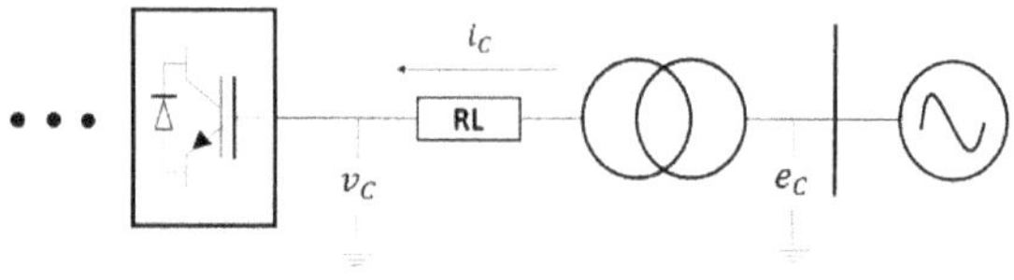

Figure 1. Voltage Source Converter single diagram.

Using the d-q transform equation, (1) can be rewritten as:

$$e_d - v_d = R_T i_d + L_T \frac{di_d}{dt} - \omega L i_q, \tag{2}$$

$$e_q - v_d = R_T i_q + L_T \frac{di_q}{dt} - \omega L i_d, \tag{3}$$

in which ω is the angular frequency of the ac voltage at PCC and the well-known power equations in d-q reference frame with $v_q = 0$ are expressed as

$$P = 3/2(v_d i_d). \tag{4}$$

$$Q = -3/2(v_d i_q). \tag{5}$$

The (4) and (5) are used in the outer controller, and the active/reactive power is controlled by the decoupled d and q axis current. The above simple equations are one of the main reasons for using the d-q current control as the fastest inner control.

2.2. Inner Current Controller of VSC

The inner current controller which a faster response than the outer controllers includes Proportional Integral (PI) controllers and makes the voltage reference. The PI controllers are used to reduce the error in the d and q axis current control with ac grid parameters. A feedforward current is used to reduce the cross-coupling effect, and feedforward voltage (v_{sdq}) is applied to compensate for the grid voltage disturbance as shown in Figure 2. Note that the system parameters depending on the feedback gain and time constant can cause problems in the inner fast control loops in a weak grid, the appropriate PI gains selection is needed. In this paper, however, selecting the optimal PI parameters is not a goal. The impact of the proposed model included in the upper hierarchical control structure is the only concern in this paper.

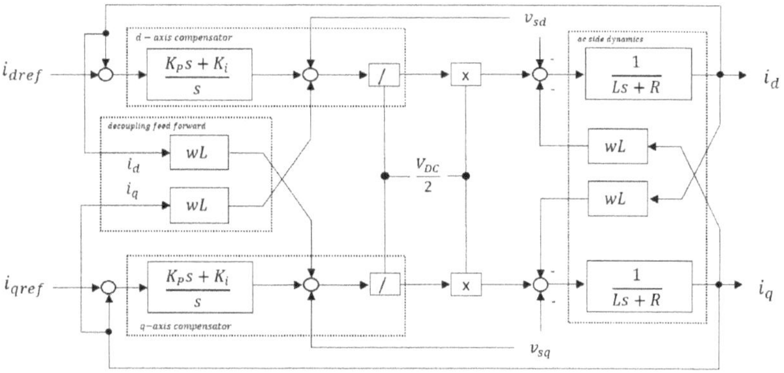

Figure 2. Inner current control loop.

Figure 2 indicates that the control plants in both the d and q axis current control loops are identical. The PI controller allows one to track the dc reference command [6]. The closed-loop transfer function that includes the ac system dynamic as L_T and R_T is represented as

$$\ell(s) = (\frac{k_p}{Ls})\frac{s + k_i/k_p}{s + R/L}, \tag{6}$$

in which the R and L are the total resistance and inductance between the PCC and the converter, respectively. Due to the system pole at $s = -R/L$, the magnitude and phase of the loop gain start to drop from a low frequency. The system pole is cancelled by the compensate zero, that is $s = -k_i/k_p$. The transfer function is rewritten by

$$G_i(s) = \frac{I_d(s)}{I_{dref}(s)} = \frac{1}{\tau_i s + 1} \cdot (k_P = \frac{L}{\tau_i}, k_i = \frac{R}{\tau_i}) \tag{7}$$

in which τ is the time constant that impacts the system response and the bandwidth of the closed loop system. Depending on the system requirements, about 0.5–5 ms is taken as an appropriate range, and the parameters of PLL and PI controller have to be decided based on grid equivalent impedance in small signal stability domain.

2.3. Outer Current Controller of VSC

Unlike the inner control loops of a fast, first-order system, outer controller makes d-q current references achieve upper control object and ensures satisfactory response [7]. The d-axis current can control the active power and dc voltage; on the other hand, the q-axis manipulates the reactive power and ac voltage through inner current loops. In the power control, the q-axis current and d-axis voltage are considered as disturbances of the d-axis current control. The specific outer controller description with the proposed models are illustrated in the next Section.

3. Novel Two Power Control Models with Communication System

3.1. The First Power Control Model of BTB VSC

The process to obtain a first power control model with the bus monitoring system is explained in this section. The goal is to make BTB VSC act like an ac transmission when contingencies occur. It contributes to the improvement of transient stability, since it reacts like other ac transmission lines that find a new convergence point after a contingency. Therefore, the ac transmission characteristics have to be involved in the BTB VSC. In the ac grid, the transfer power is determined by each side of the angle difference that provides a clue to infer the condition of the ac system. The BTB VSCs equipped with each side of bus angle information can react differently depending on the ac grid condition. To apply the angle difference variation to an active power controller, below assumption is needed [8]:

$$\Delta P = \Delta \frac{V_S \cdot V_R}{X_L} sin\delta \cong \Delta sin\delta. \tag{8}$$

The power transfer equation is illustrated as the sending end voltage V_S, receiving end voltage V_R and line impedance X_L between two buses, in which δ is the angle difference between the sending and receiving ends. With the ac voltage control, the converter constantly controls the ac voltage as 1.0 pu, and the voltage variation is assumed to be zero. The impedance that determines the amount of power is a constant value between ac grid and BTB VSC system. Finally, in the VSC, the power variation occurs when the angle difference between connected two buses is detected.

In the first power control method, the angle information of each side of ac buses are transferred through a first order lag filter, and the information used in the assistant power controller to emulate the ac transmission power flow characteristics as shown in Figure 3. The filter is required for the measured signal to remove the initial overshoots, as well as any higher angle oscillations. The power reference changes when an angle variation is detected, and the reference is obtained as

$$P^*_{total}(t) = P_{initial} + P_{res}(t - T_1) \times sin(\Delta\theta_y(t) - \Delta\theta_x(t)). \tag{9}$$

in which T_1 is a delay constant from the control lag, output limiter, and communication system. The $P_{initial}$ is initial dc power for the user-selected purpose in the steady-state condition. The $P_{initial}$ can be set for various reasons by grid operators, including the maximum power-voltage margin, improved power transmission capability, minimum power transfer cost, etc. Note that the sin function has a range of -1 to 1, the assistant power reference is always smaller than remaining active power of the converter as P_{res}. Thus, the P_{res} determines the range of the angle stability improvement.

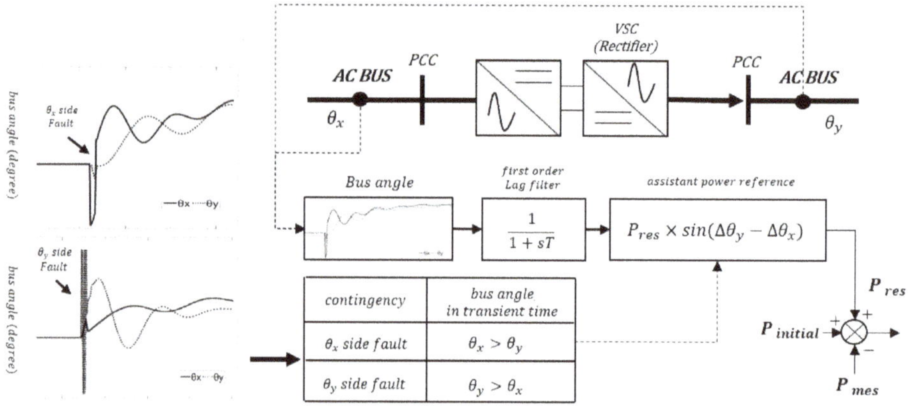

Figure 3. The first power control model structure of BTB.

There are two purposes for a set P_{res} value. First, the contingencies nearby the BTB VSC result in a high angle variation. This means that the assistant power reference could have large fluctuations in a small power system [9]. This easily impacts the system stability in which grid inertia is low. Second, the unlimited power reference could impact the system margin, leading to an unstable eigenvalue mode of active power controller. Therefore, setting the P_{res} is reasonable.

In detail, for example, when the fault occurs right next to the θ_x side of the BTB, the arbitrary bus angle at the θ_x side is immediately increased, since nearby generators increase their rotor speed to supply active power. What we want to see is the first damping of the bus angle in a sequence of a contingency. From Figure 3, the angle variation of the θ_x side more sensitively increases rather than the opposite side of θ_y. From (9), the bus angle makes $(\Delta\theta_y(t) - \Delta\theta_x(t))$ terms negative in transient time, and the P_{res} becomes negative as well. The total power output $(P_{initial} + P_{res})$ is decreased to deliver more power to the θ_x side. On the other hand, when the fault occurs right next to the θ_y side of the BTB, bus angle at the θ_y side is more increased than the other side in transient time. It makes $(\Delta\theta_y(t) - \Delta\theta_x(t))$ terms positive, and the total output power going to θ_y side is increased. In conclusion, the bus angle information reflects the ac grid condition directly after the contingency, and more power flows from the converter to the ac grid with the first power control strategy can reduce the required decelerating energy.

The operation point of the active power is consistently changed due to varying angle differences such as ac transmission line; therefore, the grid operators should accurately determine the reserve active power as P_{res}. The reason for using reserve power (P_{res}) and not using initial power $(P_{initial})$ for the proposed power control is that it can easily make the reverse direction power flow from the angle measuring system. This involves a high variation in power change, and it can worsen the ac grid condition in some cases. Also, another object of the first method is to use the remaining converter capacity to increase the converter utilization. Thus, the first power control method contribution to the stability of the ac grid is naturally constrained by a limiter.

The specific determination process of reserve active power is illustrated hereafter. The P_{res} is calculated by the converter rating (S), required reactive power (Q_{max}) from a severe contingency, and the initial active power ($P_{initial}$).

$$-\sqrt{S^2 - Q_{max}^2 - P_{initial}^2} \leq P_{res} \leq \sqrt{S^2 - Q_{max}^2 - P_{initial}^2}. \quad (10)$$

Since the ac voltage control of the embedded BTB VSC has the highest priority in the transient time, sufficient reserve capacity of the reactive power such as Q_{max} is required. However, the reactive power injecting or absorbing the ac grid can be limited by key parameters such as the ac grid voltage, converter voltage, and ac grid equivalent impedance. The maximum reactive power the converter can absorb or inject to the ac grid is limited by the two well-known equations below [10]. The reactive power is constrained by the (11) as the "critical frontier" is the relation between the transmitted active power and reactive power. Beyond the critical frontier area, the ac grid voltage becomes unstable and collapses.

$$q_{vsc} = \frac{e^2}{4x_T} - \frac{x_T}{e^2} p_c^2. \quad (11)$$

in which
q_{vsc} is the VSC reactive power;
e is the ac grid voltage;
x_T is the reactance between the VSC and the ac grid

Other equations such as (12) represent the capability of the VSC to inject the reactive power into the ac system. (12) formed circular is moved by the ac grid voltage and equivalent impedance.

$$p_{ac}^2 = \left(q_{ac} - \frac{e^2}{x_T}\right)^2 = \left(\frac{ev_c}{x_T}\right)^2. \quad (12)$$

in which
p_{ac}, q_{ac} is the apparent power at the ac system node;
v_c is the converter side voltage;

The allowable reactive power range is firstly calculated to determine the range of P_{res}. If the reactive power is not limited by the ac grid condition, the Q_{max} is chosen based on the required amount when the three-phase fault occurs near the BTB VSC system. If the reactive power consumption to maintain the voltage of the weak ac grid is large, the converter following the tendency of the ac transmission characteristic is naturally limited. The detailed determination process is illustrated below in Figure 4.

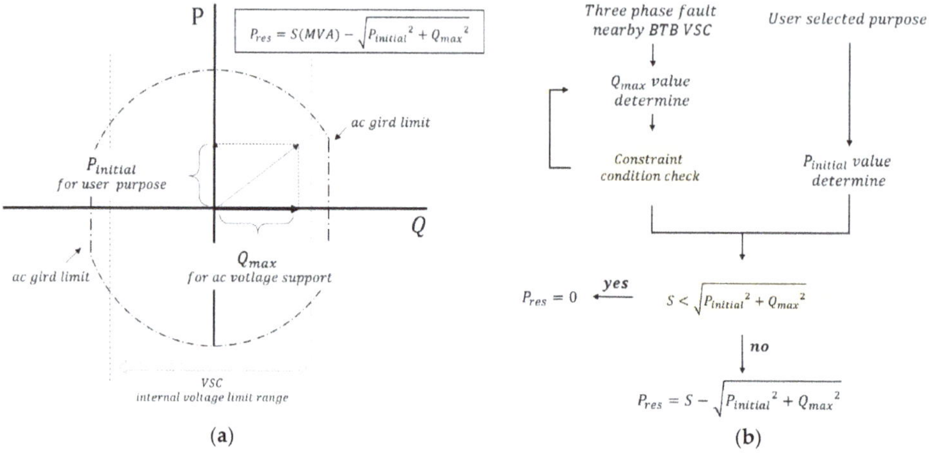

Figure 4. (**a**) Capability curve of BTB VSC and (**b**) determination process of P_{res} value.

3.2. The Second Power Control Model of BTB VSC

Each country, whether it has a different Special Protection System (SPS), as the generators tripping schemes are generally called, commonly commands specific generators to be tripped to balance the grid power. This process is generally adopted by grid operators, and to prevent the fault spread, the SPS signal that commands each generator to be tripped is activated. Under an N-2 contingency, as the level of fault increases, more generator rejection has to be applied to sustain the grid stability. The common protocol of SPS is performed according to time domain simulation with major contingencies, and the stability margin is determined by the index of acceleration and the decelerating energy [11,12].

With the second power control strategy, the flexible operation of the generator tripping scheme can be achieved without significant decelerating energy as the generators trip. The BTB VSCs surely contributes to the stability of the ac grid using a simple converter control strategy that transfers the maximum power reserve instantly to the fault area. Note that the more VSCs equipped with a proposed power control scheme the flexible tripping schemes on the generators will be possible for grid operators.

Figure 5 represents the diagram of the second power control scheme. If the SPS signal is activated, the power reference with the second power control scheme is represented by

$$P^*_{total}(t) = P_{initial} + \underbrace{[P_{res}(t - T_1) \times sin(\Delta\theta_y(t) - \Delta\theta_x(t)) \times S1]}_{The\ first\ power\ control} + \underbrace{[P_{res}(t - T_2) \times F \times S2]}_{The\ second\ power\ control}, \quad (13)$$

$$\begin{cases} F \in \{-1,\ 1\} \\ S1\ and\ S2 \in \{0,\ 1\} \end{cases}$$

in which T_2 is the delay time when the BTB VSC receives the SPS signal, and it is set by 5 cycles after the fault in this paper. The activating time is naturally faster than the generator tripping scheme as with 9 cycles, since mechanical switch is not included. The F coordinated by the grid operators only impacts the direction of the active power reserve based on the fault position, and the $S1$ and $S2$ are the switch to separate the first power control and the second power control method. The initial value of $S2$ is zero for the first method.

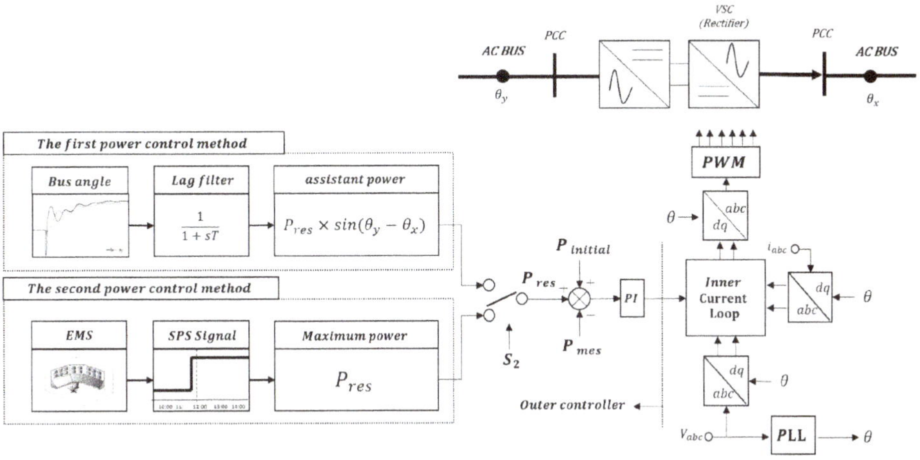

Figure 5. The second power control model of BTB VSC.

To verify the effectiveness of the second power control method, the Equal Area Criterion (EEAC) concept is introduced. From the well-established EEAC [11,12], the accelerating area is defined as A_1, and the decelerating area as A_2. To configure the stable system, the size between A_1 and A_2 has to be equal. The main cause of this improvement is that the proposed strategy can increase the height of the A_2 area during contingency. Figure 6 represents the power-angle curve and the decelerating area with and without the second power control strategy.

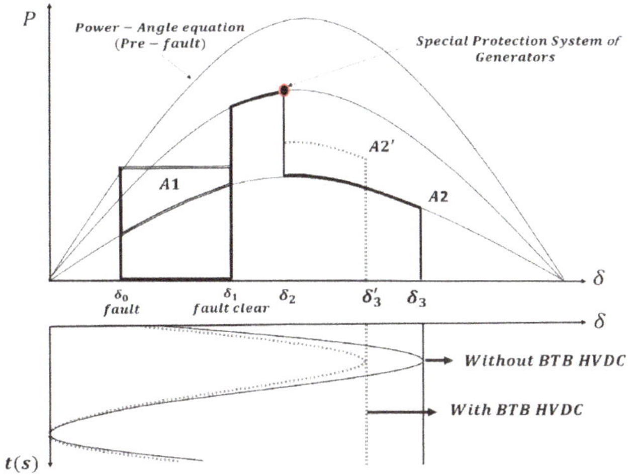

Figure 6. Equal Area Criterion with the second power control strategy.

From [11,12], the rotor angle (δ) increases by increasing the mechanical output (P_m) or by reducing the electrical output (P_e) as in (14):

$$\frac{d\delta}{dt} \propto \pm\sqrt{\int P_m - P_e d\delta}, \tag{14}$$

$$\frac{d\delta^2}{dt^2} = P_m - P_e = P_m - \frac{EVsin(\delta)}{X_T + X_e}. \tag{15}$$

in which E is an internal voltage of the generator, and X_T and X_e are the reactance of an equivalent transformer reactance and reactance of the transmission lines, respectively. The relation between A_1 and A_2 with a generation tripping scheme can be represented by (16), and the common protocols of calculating the A_2 area with the second power control scheme can be written as

$$A_2 = A_1 \times \frac{P'_m}{P_m}, \tag{16}$$

$$A_2 = \int_{\delta_2}^{\delta_3} \frac{EVsin(\delta)}{x_T + x_e} - P'_m d\delta, \tag{17}$$

$$A'_2 = \int_{\delta_2}^{\delta'_3} \frac{EVsin(\delta)}{x_T + x_e} - P'_m d\delta + \underbrace{\int_{t-T_2}^{t} P_{res}\, dt}_{\text{The second power control mehtod}} \tag{18}$$

in which $t - T_2$ is the start time when the second power controller impacts the VSCs. If the BTB VSCs receive the SPS signal where a severe contingency occurs in the ac system, the corresponding P_{res}, which is the maximum reserve power, is used. With the proposed scheme, the loss of mechanical output generated by the tripping scheme is naturally smaller than (17). (18) verifies that the first damping angle of the equivalent generators could be reduced further than the non-applied proposed operation scheme. The main one prevents the rest from a potential loss of synchronisms where the proposed scheme is applied. Eventually, the second control strategy can reduce the number of generator units ordered by SPS.

4. Simulation Study with Proposed Power Control Models

To demonstrate the effect of the proposed models, two BTB VSCs are employed to evaluate the performance of the power control structure and its impact on the behaviour of the proposed scheme under several contingencies. A simulation was carried out on the Korea power system, for which a diagram and candidate places for BTB VSCs are presented in Figure 7. Two places as 'Sinkimpo-Sinpaju' and 'SeoSeoul-Sinsungnam' have several instability network problems, including a high fault current, high angle difference, voltage instability, and overload. In order to overcome the mentioned shortcomings, KEPCO (Korea Electric Power Corporation) is considering BTB VSC systems. The detailed system specification is represented in Table 1.

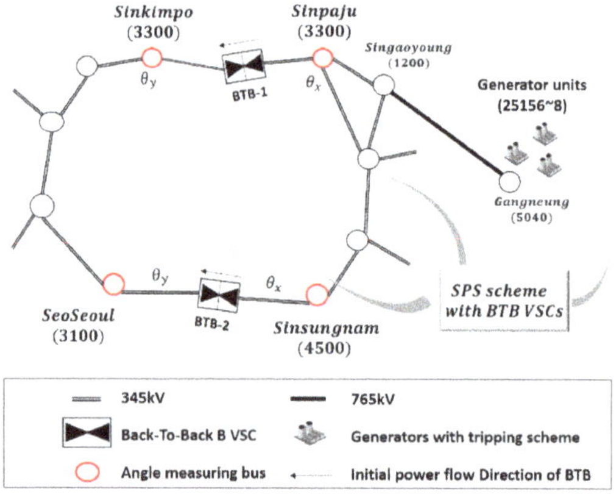

Figure 7. Simulation environment in Korea power system.

Table 1. System parameter of BTB VSC.

Item	BTB-1	BTB-2
Installation point	3300–1300	3100–4500
Rated converter	1500 MVA	1000 MVA
Equivalent impedance at BTB point	6.638 + 11.713j Ω	1.216 + j5.531 Ω
AC Terminal voltage	354 kV	
DC link voltage	400 kV	
System frequency	60 Hz	
Active power controller gain	0.8 rad/s	0.5 rad/s
Active power controller time constant	0.02 rad/s	0.01 rad/s
Inner current controller gain	1.2 rad/s	0.8 rad/s
Inner current controller time constant	0.1 rad/s	0.2 rad/s

The result is performed against the PSS/E (Power Transmission System Planning Software) with a sub-module written in the Python language. The limiter and assistant power controller that have a constant time delay from communication system were also setup on sub-module program. The initial dc power is selected to secure the maximum power-voltage margin in the ac grid, and the contingency scenarios are shown in Figure 8.

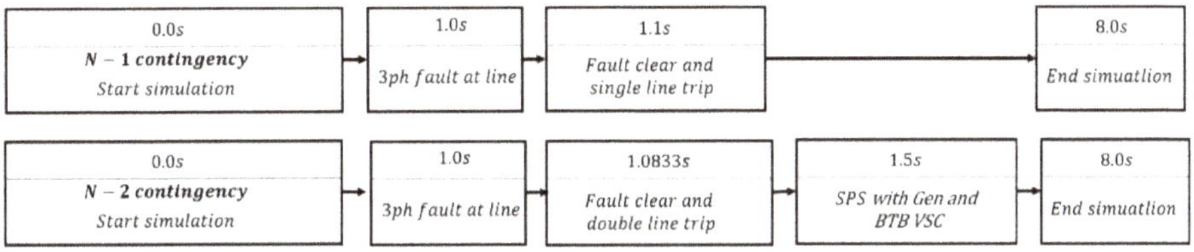

Figure 8. Contingency scenarios.

The simulation setup of the BTB VSCs is at two different places, one for the N-1 contingency at the upper place using the first power control method, and one in both places to apply the N-2 contingency with the second power control strategy. In the N-2 contingency scenario, the SPS signal is transmitted to both generators and BTB VSCs. To verify the difference between the value of the P_{res}, which is included in different limiters as "limiter-1" and "limiter-2"; the devices were stepped with the purpose of observing the contribution to the ac grid, as shown in Figure 9.

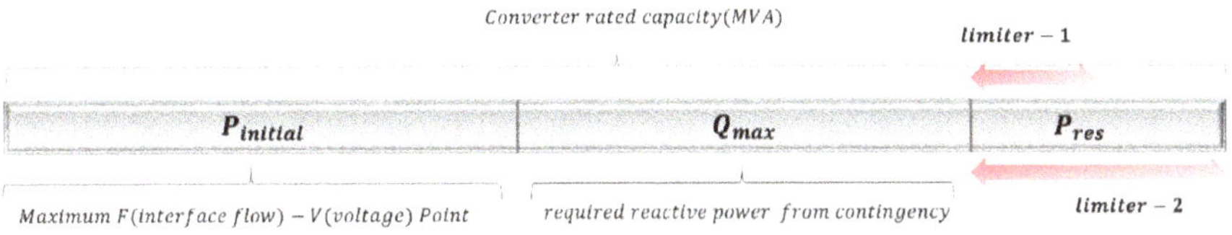

Figure 9. The limiter configuration in simulation study.

The limiter-1 has a small reserve capacity due to a large $P_{initial}$ or Q_{max} value. On the other hand, the limiter-2 allows a large reserve capacity with a small $P_{initial}$ or Q_{max} value. With the limiter-2, the BTB VSC transfers substantial active power to the fault area based on the angle difference. The reactive restrictions of the two BTB points were all guaranteed through (11) and (12).

4.1. N-1 Contingency with the First Power Control Method of BTB VSC

(1) Scenario 1—A loss of 345 kV mono-pole scenario near the Sinsungnam (4500) bus is applied to simulate the N-1 contingency, as shown in Figures 10a and 11a. A rise in the angel variation in the

fault area implies a higher command of the active power in the BTB VSC. Therefore, the first damping is mitigated at $t = 2.2s$, as shown in Figure 10a. This is because the converter reduces its power to supply more active power into the fault area that is at the θ_x side, as shown in Figure 11a.

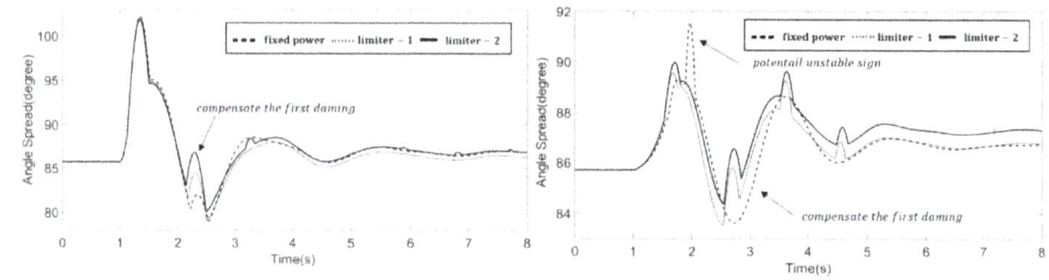

(a) (b)

Figure 10. Angle spread of ac grid, (**a**) left side fault (θ_y), and (**b**) right side fault (θ_x).

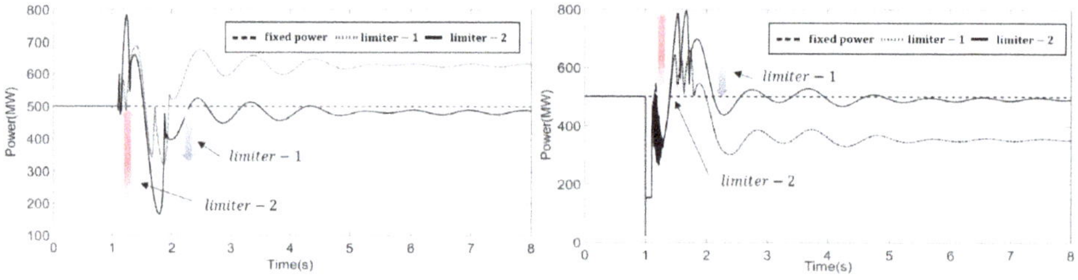

(a) (b)

Figure 11. Active power of BTB VSCs, (**a**) left side fault (θ_y), and (**b**) right side fault (θ_x).

(2) Scenario 2—Contrary to the first scenario, a loss of a 345kV mono-pole near the SeoSeoul (3100) bus is discussed in Figures 10b and 11b. The fixed power control, illustrated in Figure 10b, has a potentially unstable condition at $t = 1.9s$ due to the first damping. The stabilization of this situation requires several types of SPS to balance the power. For a VSC under the first control scheme, the first angle damping can be mitigated, leading to a stable system. Again, given the different limiter, the bold line as limiter-2 has a large capacity for active power compared to limiter-1. The value of P_{res} increased, and the contribution to the angle stability increased. The findings from Figures 10 and 11 indicate that a trade-off must be made between improvements in the angle stability and the economics related to the convert sizing. Also, the power control in BTB VSC is demonstrated to contribute to the ac grid, besides having other advantages.

In addition to the above cases and for the sake of clarity, the result of three more contingencies using the first power control model is illustrated in Table 2. Note that each contingency made a different angle variation, the effects of the applied proposed scheme are all different.

Table 2. The effect of the first power control scheme in other contingencies.

Contingency Num	Fault Location	Improvement Range of First Angle Damping
1	1550–1400	6°
2	1350–1301	2.6°
3	1400–1800	4.5°

4.2. N-2 Contingency with the First Power Control Method of BTB VSC

In Korean power systems, to satisfy the grid reliability, the SPS was triggered by the generator tripping scheme. To verify the second power control with the SPS signal, a simulation with a loss of a 765 kV double-pole is applied in this section. During the contingency, the three generators that are represented by Table 3 must be tripped to maintain the power balance.

Table 3. The generators' status-receiving SPS signal.

Gen Num	Gen Name	Gen Capacity
25156	Hanul #6	1 GW
25157	Hanul #7	1.5 GW
25158	Hanul #8	1.5 GW

Unlike the present situation in Korean grid operation systems, the two BTB VSCs also receive the signal of the SPS with other generators only in this scenario. The second power control model sets their maximum power in the BTB VSC according to (13). Figure 12 represents the active power of two BTB VSCs according to two different limiters. The fixed power control with BTB VSCs makes the power system unstable with a two-generator tripping scheme, as in Figure 13. However, from the angle stability analysis with a time domain simulation, this result demonstrates that using the second power control scheme with BTB VSCs could reduce the tripped generator from three to two as shown in Table 4, and the stability margin is adequately acquired. Through this simulation, it was demonstrated that the BTB VSCs with SPS signal can be applied to a commensurate protection system in a large interconnected system.

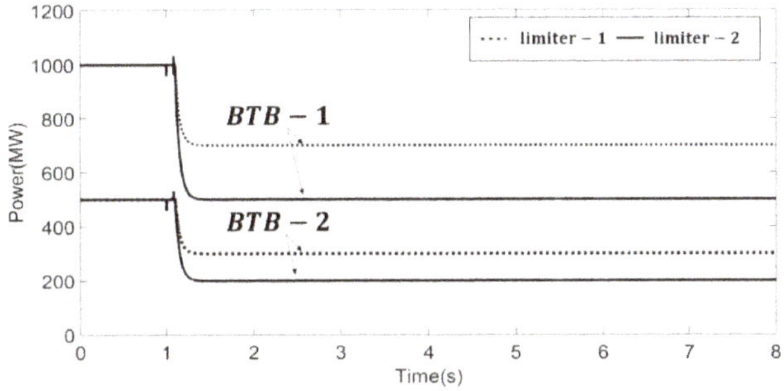

Figure 12. Active power of BTB VSCs with different P_{res}.

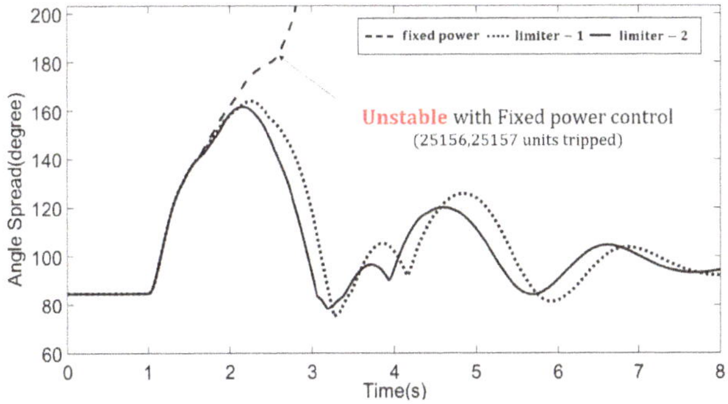

Figure 13. Angle spread result in an N-2 Contingency.

Table 4. Reduced tripped generator with the second power control scheme in N-2 contingency.

	Without the Second Power Control	With the Second Power Control
Tripped Generators	25156; 25157; 25158	25156; 25157

5. Discussion

In the complex ac grid, the grid operators find it hard to predict when the contingencies will occur. Furthermore, changing the control mode depending on each fault scenario is annoying. To resolve such problems, BTB VSCs with two power control strategies are proposed in this paper, and their capability for angle stability was sufficiently demonstrated in an unpredicted contingency. The proposed control schemes are also able to act like a form of SPS, and depending on the contingency scenario, the effectiveness of the proposed models can be more escalated.

6. Conclusions

The BTB VSC is commonly installed to increase the voltage stability of a weak ac grid or the interconnection points of renewable systems. Also, it plays an apparent role in reducing the fault current magnitude and increasing the transfer capability in the ac transmission. In this paper, however, to maximize the advantages of embedded BTB VSCs, new power control strategies are proposed to improve the angle stability. The novel power control models are demonstrated to easily contribute angle stability in the ac grid with a simple assistant controller while reducing the number of tripped generator units.

Author Contributions: The main control schemes were proposed by S.S and G.J.; The experiment results were collected and analyzed by S.H, B.K and S.C.

Funding: This research received no external funding

Acknowledgments: This work was supported under the framework of the international cooperation program managed by the National Research Foundation of Korea (No. 2017K1A4A3013579) and the Human Resources Program in Energy Technology of the Korea Institute of Energy Technology Evaluation and Planning (KETEP).

Conflicts of Interest: The authors declare no conflict of interest.

References

1. Kidd, D.; Mehraban, B.; Ekehov, B.; Ulleryd, J.; Edris, A. Eagle pass back to back VSC installation and operation. In Proceedings of the Power Engineering Society General Meeting, Portland, OR, USA, 5–9 August 2018; pp. 1829–1833.
2. Sankar, S.; Rector, J.; Fairly, P.; Marz, M.; Copp, K.; Manty, A.; Irwin, G. ATC's Mackinac Back-to-Back HVDC Project: Planning and Operation Considerations for Michigan's Eastern Upper and Northern Lower Peninsulas. In *CIGRE US National Committee 2013 Grid of the Future Symposium*; ISO New England Inc.: Holyoke, MA, USA, 2013.
3. Pan, J.; Nuqui, R.; Srivastava, K.; Jonsson, T.; Holmberg, P.; Hafner, Y.-J. AC grid with embedded VSC-HVDC for secure and efficient power delivery. In Proceedings of the Energy 2030 Conference, Atlanta, GA, USA, 17–18 November 2008; pp. 1–6.
4. Kim, M.-Y.; Song, Y.-U. The Analysis of Active Power Control Requirements in the Selected Grid Codes for Wind Farm. *J. Electr. Eng. Technol.* **2015**, *10*, 1408–1414. [CrossRef]
5. Fang, X.; Chow, J.H. BTB DC link modeling, control, and application in the segmentation of AC interconnections. In Proceedings of the Power & Energy Society General Meeting, Calgary, AB, Canada, 26–30 July 2009; pp. 1–7.
6. Yazdani, A.; Iravani, R. *Voltage-Sourced Converters in Power Systems: Modeling, Control, and Applications*; John Wiley & Sons: Hoboken, NJ, USA, 2010.
7. Rouzbehi, K.; Miranian, A.; Candela, J.I.; Luna, A.; Rodriguez, P. A generalized voltage droop strategy for control of multiterminal DC grids. *IEEE Trans. Ind. Appl.* **2015**, *51*, 607–618. [CrossRef]

8. Song, S.; Kim, J.; Lee, J.; Jang, G. AC Transmission Emulation Control Strategies for the BTB VSC HVDC System in the Metropolitan Area of Seoul. *Energies* **2017**, *10*, 1143. [CrossRef]
9. Konishi, H.; Takahashi, C.; Kishibe, H.; Sato, H. A consideration of stable operating power limits in VSC-HVDC systems. In Proceedings of the Seventh International Conference on AC and DC Transmission, London, UK, 28–30 November 2001; pp. 102–106.
10. Pinto, R.T. Multi-Terminal DC Networks: System Integration, Dynamics and Control. Ph.D. Thesis, Delft University of Technology, Delft, The Netherlands, 4 March 2014.
11. Xue, Y.; Wehenkel, L.; Belhomme, R.; Rousseaux, P.; Pavella, M.; Euxibie, E.; Heilbronn, B.; Lesigne, J.-F. Extended equal area criterion revisited (EHV power systems). *IEEE Trans. Power Syst.* **1992**, *7*, 1012–1022. [CrossRef]
12. Xue, Y.; Van Custem, T.; Ribbens-Pavella, M. Extended equal area criterion justifications, generalizations, applications. *IEEE Trans. Power Syst.* **1989**, *4*, 44–52. [CrossRef]

applied sciences

MDPI

Article

Development of A Loss Minimization Based Operation Strategy for Embedded BTB VSC HVDC

Jaehyeong Lee [1], Minhan Yoon [2], Sungchul Hwang [1], Soseul Jeong [1], Seungmin Jung [3] and Gilsoo Jang [1,*]

[1] School of Electrical Engineering, Korea University, Anam-ro, Sungbuk-gu, Seoul 02841, Korea; bluesky6774@korea.ac.kr (J.L.); adidas@korea.ac.kr (S.H.); jss928@korea.ac.kr (S.J.)
[2] Department of Electrical Engineering, Tongmyong University, 428, Sinseon-ro, Nam-gu, Busan 48520, Korea; minhan.yoon@gmail.com
[3] Department of Electrical Engineering, Hanbat National University, Daejeon 305-719, Korea; seungminj@hanbat.ac.kr
* Correspondence: gjang@korea.ac.kr; Tel.: +82-010-3412-2605

Received: 4 April 2019; Accepted: 28 May 2019; Published: 30 May 2019

Abstract: Recently, there have been many cases in which direct current (DC) facilities have been placed in alternating current (AC) systems for various reasons. In particular, in Korea, studies are being conducted to install a back-to-back (BTB) voltage-sourced converter (VSC) high-voltage direct current (HVDC) to solve the fault current problem of the meshed system, and discussions on how to operate it have been made accordingly. It is possible to provide grid services such as minimizing grid loss by changing the HVDC operating point, but it also may violate reliability standards without proper HVDC operation according to the system condition. Especially, unlike the AC system, DC may adversely affect the AC system because the operating point does not change even after a disturbance has occurred, so strategies to change the operating point after the contingency are required. In this paper, a method for finding the operating point of embedded HVDC that minimizes losses within the range of compliance with the reliability criterion is proposed. We use the Power Transfer Distribution Factor (PTDF) to reduce the number of buses to be monitored during HVDC control, reduce unnecessary checks, and determine the setpoints for the active/reactive power of the HVDC through system total loss minimization (STLM) control to search for the minimum loss point using Powell's direct set. We also propose an algorithm to search for the operating point that minimizes the loss automatically and solves the overload occurring in an emergency through security-constrained loss minimization (SCLM) control. To verify the feasibility of the algorithm, we conducted a case study using an actual Korean power system and verified the effect of systematic loss reduction and overload relief in a contingency. The simulations are conducted by a commercial power system analysis tool, Power System Simulator for Engineering (PSS/E).

Keywords: back-to-back HVDC; embedded HVDC; HVDC operation point; Powell's direct set method; system loss minimization

1. Introduction

In recent years, the electric-power industry has changed from a conventional alternating current (AC) system to a hybrid system, in which special facilities such as high voltage direct current (HVDC) and flexible AC transmission system (FACTS) work together. In particular, HVDC has been used for over 50 years, mainly for asynchronous interconnection, long-distance transmission, submarine, and underground cable transmission, etc. The most commonly used system is the line-commutated converter (LCC) HVDC, which uses thyristor valves, but the development of voltage-sourced converter (VSC) HVDC technology is also on the rise with the advances in power electronics technology [1–3].

In general, HVDC is mainly used with point-to-point (PTP) type which is used for large-capacity, long-distance transmission. These HVDCs are intended to transmit power unidirectionally between the two regions, so no special operation strategies are required. Recently, however, small-and medium-scale VSC HVDCs were installed in the grid to provide additional grid services in addition to unidirectional power transmission, because of the independent control of active/reactive power and converter switching operation of the VSC HVDC, which can provide grid services for an AC system, as shown in Table 1 [2,4].

Table 1. Grid services of voltage-sourced converter (VSC) high voltage direct current (HVDC) for alternating current (AC) systems.

Control and Support	Grid Service
Active power control and frequency support	- Primary control or frequency containment reserves (FCR) - Secondary control or frequency restoration reserves (FRR) - Tertiary control or replacement reserves (RR)
Reactive power control and voltage support	- Reactive power absorb/supply
Rotor angle stability-related control	- Avoiding loss of synchronism - Damping electromechanical oscillations
Other	- Power oscillation damping capability - DC power flow control - Black start capability - Loss compensation - DC transmission reserve

In particular, it is easy to provide grid services for the AC grid when embedded HVDC is installed in the grid, where embedded HVDC is defined as "a DC link with multiple ends that are physically connected from a single synchronous AC network. With such a connection, these connections allow DC to perform basic functions of bulk power transmission as well as some additional control functions within the AC network such as power flow control, voltage control, stability improvement" [5]. In other words, embedded HVDC is mainly used for AC power flow control by active power control at both ends of the converter, improvement of voltage stability by reactive power control, suppression of failure propagation in the AC system, and system stability through frequency control. With the introduction of embedded VSC HVDC, there have been various studies of embedded VSC HVDC operation strategies as shown in [6–21]. References [6–8] are early research on the operation strategy of the HVDC, which improves the stability of power facilities by Remedial Action Scheme (RAS). As defined [9,10], an RAS is designed to take corrective action to maintain system reliability and provide acceptable system performance. This operating strategy has been applied to many HVDC projects such as Manitoba HVDC, Eastern/Western Alberta Transmission Line, and the RAS is used as the basis of emergency control in this paper [11–16]. The operating point of HVDC should be determined prior to these emergency controls, and most HVDCs are intended for long-distance transmission, so the operating point is determined according to the rated capacity or N-1 failure without any special strategy [17]. However, by changing the HVDC operating point, system total loss can be reduced, and such studies have also been conducted. This method mainly focuses on reducing transmission loss with Multi-Terminal DC (MTDC) [18,19], and studies on finding operating points based on objective functions such as system loss, voltage deviation, and reliability criteria have been variously conducted on small scale systems [20,21]. These conventional studies have the drawback such as they were conducted in a small-scale radial system, not a large-scale meshed system; the studies manly consider only AC loss; and there is no integrated control scheme for HVDC operation considering system total loss.

In this paper, a method for determining the normal/contingency operation point of embedded back-to-back (BTB) VSC HVDC is proposed. In calculating the operation point during normal conditions, the system total loss minimization (STLM) control to minimize total losses within the grid reliability criterion is conducted using Powell's direct set theory. At this time, by using the power transfer distribution factor (PTDF), we calculate the lines that are sensitive to HVDC operation beforehand, reduce the number of lines to be monitored, and exclude the investigation of unnecessary lines. In addition, the security-constrained loss minimization (SCLM) control for eliminating the overloads in N-1 contingencies operates additionally. The optimum HVDC operating point is determined by setting the system total loss reduction as the objective function even within the operating range in which the overload can be eliminated by a method different from the existing remedial control. This paper is structured as follows. Section 2 describes the Korean power system briefly and provides a concept of BTB VSC HVDC in a meshed system. The method for finding the operating point of embedded BTB VSC HVDC is described in Section 3. Section 4 shows the simulation results of the BTB VSC HVDC operating point decision process for the Korean power system and demonstrates the validity of the proposed algorithm. The simulations are conducted by Power System Simulator for Engineering (PSS/E) 33.4.0 and the optimization results are implemented in Python code.

2. System Description

2.1. Korean Power System

In South Korea, 40% of the total load is concentrated in the metropolitan area, so the system is composed of multi-looped types for improving system stability. This system is advantageous for most aspects of stability, but it makes the fault current problem more serious. In order to solve the fault current problem, current limiting reactor installation, increasing circuit breakers (CB), grid segmentation, and other suggestions have been proposed. However, none of the methods can solve the fault current problem ultimately. Recently, as an alternative to these, a grid segmentation using BTB HVDC has been proposed. In this aspect, fault current reduction can be cited as a criterion for deciding on the HVDC position in the meshed power system. Currently, two long-distance HVDCs are in operation in the Korean power system, and four HVDCs will be added in the future. One of them is an embedded-type HVDC, which will be installed in the Seoul metropolitan area to solve the fault current problem. Studies on the installation of BTB HVDC for resolving the fault current problem in the metropolitan area have been actively conducted, as shown in [22,23].

2.2. Embedded BTB HVDC Siting in Meshed System

When replacing the existing AC line of the system with HVDC, the equivalent impedance of the power system increases, and the magnitude of the fault current decreases accordingly. For short circuit analysis, HVDC normally operates as a PQ bus at both ends for load flow calculation, whereas the fault current mode acts as if the current source is connected to both ends as shown in Figure 1 [24].

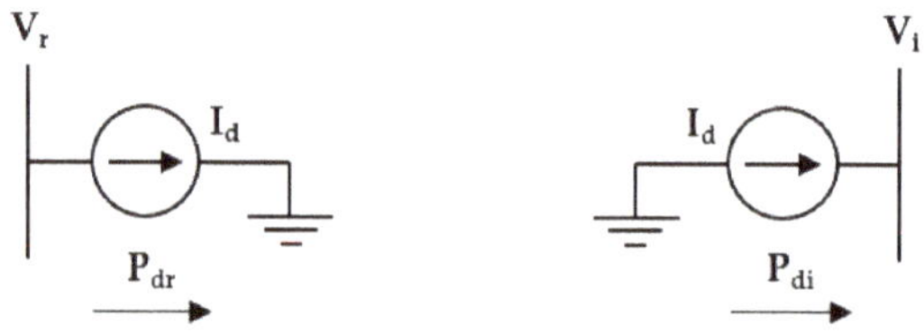

Figure 1. HVDC fault calculation model (PQ bus mode).

The three-phase fault current is calculated as the sum of the system Thevenin impedance of the fault point and the fault impedance formed between the fault point and the ground divided by the voltage that was applied just before the fault occurred. In other words, it is represented by a model in which an open circuit voltage source having a voltage applied to a point just before a failure of a short circuit formed at the time of failure is connected as in Equation (1):

$$I^n_{fault} = \frac{V^n_{prefault}}{Z^n_{Th} + Z^n_{fault}}, \tag{1}$$

in which

I^n_{fault} is a fault current at the fault point n;
$V^n_{prefault}$ is a pre-fault voltage at the fault point n;
Z^n_{Th} is a thevenin impedance at the fault point n; and
Z^n_{fault} is a fault impedance at the fault point n;

In Equation (1), $V^n_{prefault}$ and Z^n_{fault} cannot be controlled arbitrarily, and when the AC line is replaced with BTB HVDC, Z^n_{Th} can be controlled according to the position. Studies have been conducted to find the position of HVDC to reduce the fault current caused by the increase of Z^n_{Th}. In this paper, using the HVDC decision algorithm defined in [23], the HVDC placement is chosen. Although it is possible to use the location selection criteria to reduce the fault current of the entire system or to reduce the number of fault current exceeding points, the location of HVDC is selected based on the fault current reduction of the most severe buses.

2.3. Proposed Control Scheme

Figure 2 briefly shows a block diagram of the BTB HVDC operating strategy proposed in this paper. The proposed HVDC operating point decision method can be divided into two major components. The first is the operating point in the steady state. It is the point that can minimize the loss of the entire system within the range of the system reliability criterion. In order to minimize the system total losses, AC voltage control at both ends of the HVDC is performed for proper reactive power output as well as active power control. The second is the operation strategy of HVDC at the time of disturbance occurrence. Unlike AC systems, HVDC produces a constant output regardless of system conditions without special control. This feature can cause overloads on other branches and can also have a detrimental effect in terms of losses, so an emergency control strategy of embedded HVDC is essential. In case of the emergency control, the operation point is changed to the range which eliminates the overload of the branches, and then the point is controlled to the operation point which can minimize the total losses at that system condition.

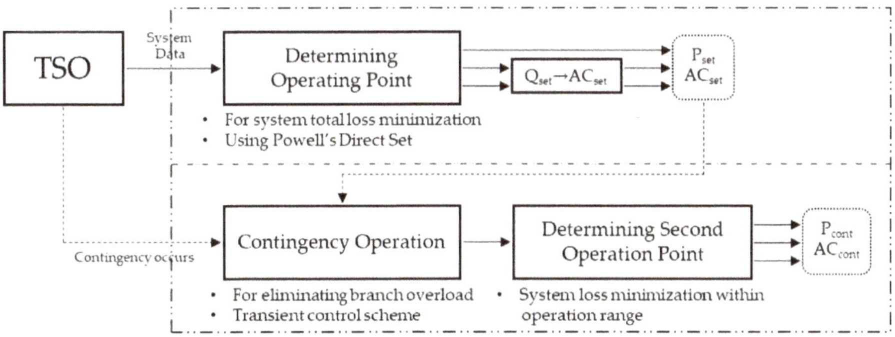

Figure 2. Proposed control scheme.

3. Determination of Embedded HVDC Operation Point

3.1. System Total-Loss Minimization (STLM) Control

In this paper, we consider the following points when determining the HVDC operating point at steady-state. After setting the constraints on each point, the optimum operating point of the embedded HVDC is calculated using the optimization method.

- System total-loss minimization
- Voltage criteria
- Branch overloads

3.1.1. System Total-Loss Minimization

The most important factor that affects the determination of the operating point is the minimization of the total loss of the system. The most important advantage of VSC HVDC is independent control of active power and reactive power, which can be used to change the direction of the system's power flow, leading to a change in system losses. Therefore, in this study, we decided on the operating point that minimizes the system's total loss within the capacity of HVDC. The system's total loss means the sum of the existing AC loss and HVDC loss.

$$\begin{aligned} P_{loss,total} &= \sum_{n_l} P_{loss,AC} + \sum_{n_{l,DC}} (P_{loss,DC} + P_{loss,conv}) \\ &= \sum_{n_l} \frac{P^2_{r,\,AC} + Q^2_r}{V^2_{r,AC}}, R_{AC} + \sum_{n_{l,DC}} \left(\frac{P^2_{r,\,DC}}{V^2_{r,DC}}, R_{DC} + (P_{r,DC} + P_{s,DC}), \lambda \right) \end{aligned} \tag{2}$$

in which

$P_{loss,total}$ is the system total loss;
$P_{loss,AC}$ is the AC system loss;
$P_{loss,DC}$ is the HVDC line loss;
$P_{loss,conv}$ is the HVDC conversion loss;
$P_{r,\,AC}$, $P_{r,DC}$ are the AC/DC line flow active powers at the receiving end;
$P_{s,DC}$ is the DC line flow active powers at the sending end;
Q_r is the AC line flow reactive powers at the receiving end;
$V_{r,AC}$, $V_{r,DC}$ are the AC/DC voltage at the receiving end;
R_{AC}, R_{DC} are the resistance values of the AC/DC line; and
λ is the loss rate of the converter.

The VSC DC losses have been decreasing over the years because of the application of Modular Multi-Level Converter (MMC) topology and optimization of switching logic. For example, the HVDC VSC loss of the fourth-generation technology is now comparable with the HVDC LCC ("Classic") technology [25,26]. In this paper, conversion loss is considered to be 0.9% of the active power setpoint per converter, but the value of the technology can be lowered according to development.

In addition, since BTB HVDC has no line loss, the DC related losses did not include the DC line loss, but only the conversion loss of the converter, which has a big influence on the loss reduction of the entire system. Therefore, Equation (2) is simplified as in Equation (3), and from the viewpoint of loss reduction, the conversion loss of BTB HVDC can greatly influence the numerical value of the entire system. At this time, since the converter and the inverter are at the same position, it is assumed that $P_{r,DC} = P_{s,DC}$.

$$\text{Minimize } P_{loss,total} = \sum_{n_l} \frac{P^2_{r,\,AC} + Q^2_r}{V^2_{r,AC}}, R_{AC} + \sum_{n_{l,DC}} 2, P_{r,DC}, \lambda. \tag{3}$$

3.1.2. Voltage Criteria

The active/reactive power control of the BTB VSC HVDC can have a significant effect on the voltage of the near buses. Generally, embedded HVDC operates in *AC* voltage control mode. However, in this paper, we decided on the proper operating point in the reactive power setpoint control mode to maximize loss reduction and operate in *AC* voltage control mode to maintain the operating point.

$$\underline{V_i} \leq V_i \leq \overline{V_i} \tag{4}$$

in which

$\underline{V_i}$, $\overline{V_i}$ is the minimum/maximum voltage at bus i; and
V_i is voltage at bus i.

3.1.3. Line Overload

The active/reactive power control of HVDC also affects the power flow of nearby lines. In this paper, we restrict the operation of BTB VSC HVDC so that overloads do not occur on other lines in the steady state.

$$\underline{F_{line,l}} \leq F_{line,l} \leq \overline{F_{line,l}} \tag{5}$$

in which

$\underline{F_{line,l}}$, $\overline{F_{line,l}}$ is the minimum/maximum power flow on line l; and
$F_{line,l}$ is power flow on line l.

The change of the BTB VSC HVDC operating point has a significant effect on some lines, but not all lines. Monitoring the overload of all the lines depending on the system conditions is a burden of communication. Therefore, it is necessary not to observe all the lines, but to find lines that are sensitive to the operation of the HVDC. At this time, the bus lines to be observed are chosen using the PTDF described in [27–29]. PTDF is an index of line sensitivity based on the DC power flow model, and it is easy to analyze system reconfiguration and installation of DC facilities. When the power to the specific bus changes, it is possible to determine the sensitivity of the lines by indicating the influence of the power flow change on the lines. The PTDF is expressed as shown in Equation (6).

$$\varphi_m^l = \frac{\Delta F_l}{\Delta P_m} \tag{6}$$

in which

φ_m^l is the PTDF of line l with respect to bus m;
ΔF_l is the amount of the power flow change in line l; and
ΔP_m is the amount of the power change injected for bus m.

The *AC* line sensitivity caused by HVDC can be represented by the PTDF values at both ends and capacity of the HVDC as shown in Figure 3. For the HVDC operation, PTDF values at both ends of the HVDC are similar for the *AC* lines where the power change is not large (low sensitivity); and in the opposite case, the difference between the PTDF values is large, and the result of Equation (7) is large too. Equation (7) is obtained by multiplying the difference of the PTDF values at the end of the HVDC for any *AC* line by the capacity of the HVDC, and it predicts the maximum value that can be reached in the *AC* line when the operation of the HVDC is maximized. The obtained value can be compared with the reference value of each line to decide whether to consider the *AC* line in the monitoring line or not. The reference value is obtained by weighting the capacity of each line as shown in Equation (8). In Equation (8), W affects the number of monitoring lines. The larger the W value, the smaller the number of monitoring lines, and the smaller the value of W, the larger the number of monitoring lines. Therefore, even when the W value is large, the branches determined as the monitoring line can be regarded as a line that is highly influenced by the HVDC operation. On the contrary, when W is controlled to a lower value, relatively less sensitive lines are considered as monitoring branches, so that it is possible to operate the system in detail, but it can be a heavy burden for HVDC operators. The system operator should carefully determine the W value according to the purpose and convenience.

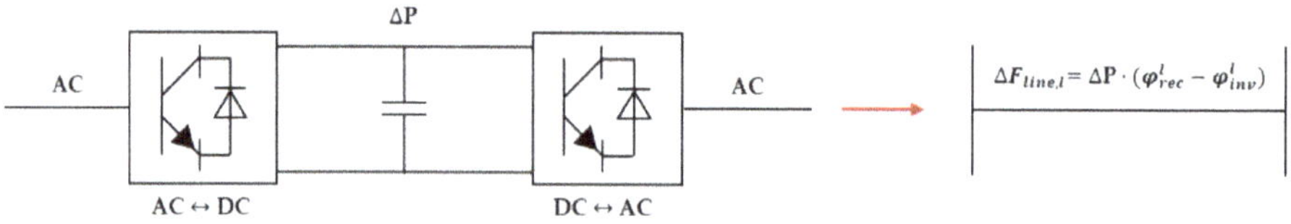

Figure 3. Power flow change caused by back-to-back (BTB) HVDC active power operation.

$$M^l_{normal} = 2 \cdot P_{cap}, \left(\varphi^l_{rec} - \varphi^l_{inv}\right), \tag{7}$$

$$M^l_{ref} = S^l_{MVA} \times W, \tag{8}$$

$$if\ M^l_{normal} \geq M^l_{ref}\ :\ \text{Choose line l as monitoring line}, \tag{9}$$

$\Delta F_{line,l}$ is the power flow change of line l by HVDC control
M^l_{normal} is the maximum variation of line l;
M^l_{ref} is the reference value of line l;
P_{cap} is the active power capacity of the HVDC;
φ^l_{rec} is the PTDF of line l with respect to the rectifier end;
φ^l_{inv} is the PTDF of line l with respect to the inverter end;
S^l_{MVA} is the capacity of line l; and
W is a weighting factor.

The monitoring lines determined using the PTDF should consider both normal and contingency conditions. Especially, in contingency, when the line is tripped after the accident, PTDF is calculated again, because the system topology changed. Therefore, it is necessary to pre-calculate the PTDF after the line is tripped for the main contingencies, and the sensitive lines should also be determined as the monitoring line.

3.1.4. Application of Three-Dimensional Powell's Direct Set (3-PDS)

The operating point of the BTB VSC HVDC is found from the minimization of the total loss (Equation (3)), the voltage reference (Equation (4)), and the thermal limit (Equation (5)) as described above. In this paper, Powell Direct Set (PDS) theory was used in an optimization method to find the operating point [30]. The PDS method can directly search for optimum point, instead of using differential, and obtains the optimization point using its direction. Therefore, it is easy to apply and is not restricted by whether it can be differentiated [31,32].

Especially, VSC HVDC uses three-dimensional PDS because it has three controllable elements: $\{P,\ Q_{rec}, Q_{rec}\}$, that are independent of each other. In order to find the three-dimensional minimum point, we tried to find the operating point decision point of HVDC more easily by using 3-PDS instead of finding the differentiation point. The main steps to determine the point of operation using the 3-PDS method are shown as follows.

1. Set the initial point $S_0^{(1)}$ and the independent directions used for initial searching ϵ. We set the values as $S_0^{(1)} = (0, 00)$, $\epsilon = \{\ P\ \rightarrow,\ \overrightarrow{Q_{rec}}, \overrightarrow{Q_{inv}}\}$.

2. A conjugate direction should be generated within each iteration.

   - Starting from $S_0^{(k)}$ (k is the iteration number), sequentially search the directions in ϵ by finding the minimum.

$$Min\ F_{obj}\left(S_i^{(k)}\right) = F_{obj}\left(S_{i-1}^{(k)} + \delta\Delta\epsilon\right), \tag{10}$$

   where δ is the step size.

   - A conjugate direction $\epsilon_{conju.}^{(k)}$ is generated after searching down by

$$\epsilon_{conju.}^{(k)} = S_i^{(k)} - S_0^{(k)}. \tag{11}$$

3. Update the search direction ϵ by adding directions $\epsilon_{conju.}^{(k)}$ to ϵ and replace the other direction. It is common to replace the first direction with a new $\epsilon_{conju.}^{(k)}$, but in this application, since the control of active power is much more influential than the control of reactive power in loss reduction, so the direction of active power (P $\rightarrow$) is left and the new direction is updated. We can get $\epsilon = \{\ P\ \rightarrow, \epsilon_{conju.}^{(k)}, \epsilon_{conju.}^{(k+1)}\}$.
4. Find the new optimal operation point $S_0^{(k+1)}$.
5. Convergence check.

3.1.5. HVDC Operating Point with STLM Control

In this paper, the optimum operating point of HVDC is calculated by using Equations (4)–(9) as the constraint condition and Equation (3) as the objective function. Using the STLM control, we can establish the active/reactive power operation point of HVDC that minimizes total system loss by using the algorithm of Figure 4 and, in contingencies, apply other control strategies. The explanation of Figure 4 as follows.

1. The initial active/reactive power operation range is set with reference to the HVDC capacity.
2. The number of branches to be monitored could be reduced using PTDF. These represent branches where the amount of power flow varies greatly with HVDC operating point (Section 3.1.3).
3. The operating area should be adjusted so as not to deviate from the constraints. In the later operating point determination process, the operating range set here should not be exceeded.
4. The operation point that minimizes the system total loss is determined using the PDS. The active/reactive power output determined in this process are respectively input into the active power set value and the AC voltage set value.

3.2. *Security-Constrained Loss Minimization (SCLM) Control*

When a disturbance occurs in the system, line and transformer overloads occur in the system. In order to eliminate the overloads, it is necessary to control the operating point of the HVDC, which is called a "security-constrained loss minimization (SCLM) control." The action is as follows.

1. BTB VSC HVDC normally operates with the operating point calculated in Section 3.1.
2. In the event of a contingency, it detects the overloads of the other line and calculates the operation range where the overload of the line can be eliminated.
3. Within the previously determined operating area, a new operating point is calculated that minimizes losses and maintains system reliability.
4. At this time, it is necessary to check whether the other lines are overloaded at the determined operating point. If there are overloads of other lines, an optimum operating point should be found in the operating area where the overloads of the line are eliminated.

Figure 5 shows an algorithm of the SCLM control in the event of a contingency.

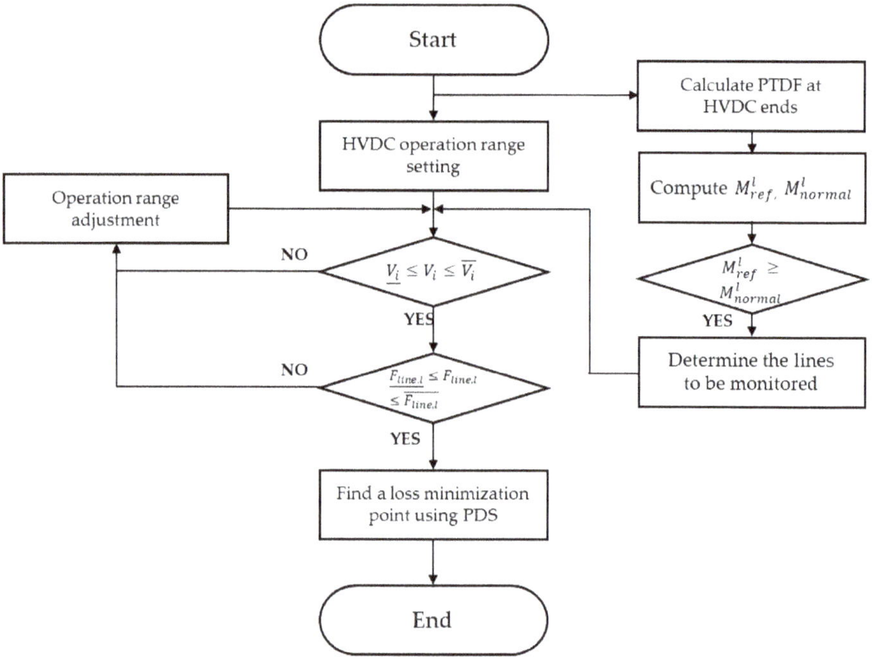

Figure 4. Algorithm of the system total loss minimization (STLM) control.

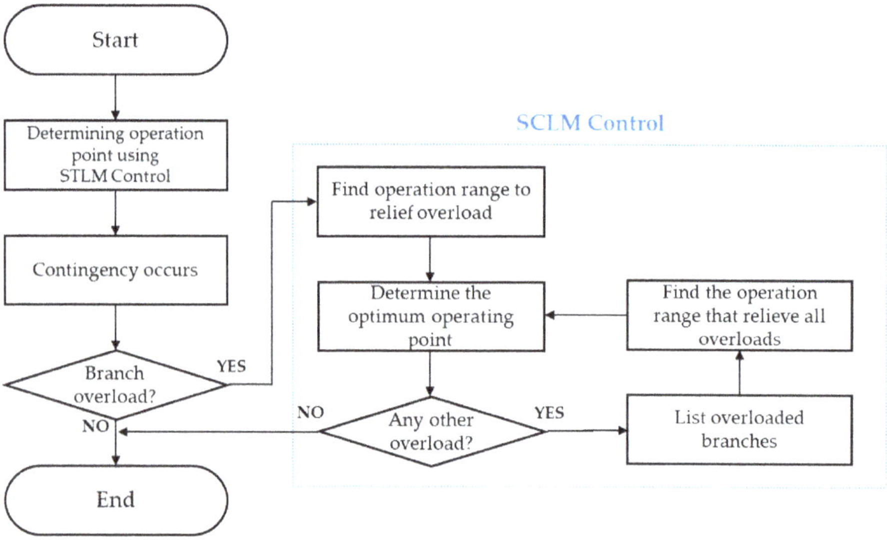

Figure 5. Algorithm of the security-constrained loss minimization (SCLM) control.

4. Case Study

4.1. System Information

For case studies, we used the Korean metropolitan power system in 2021, the features of which are as follows.

- The load in the metropolitan area is about 40% of the total load, and a large amount of power is supplied from the non-metropolitan area.
- In order to improve the system stability, it is configured as a looped system, and thereby the problem of fault current is serious.
- The 154kV Yangju bus (1410) will be separated and connected to VSC BTB HVDC to solve the fault current problems in the metropolitan area and to improve system stability.
- Detailed information on the system is shown in Table 2; detailed specifications and system configuration of HVDC are shown in Table 3 and Figure 6.

Table 2. System information.

Item	Information	
Case data	Korea Power System (2021)	
System total generation	98,771.0 MW	
System total load	97,200.8 MW	
Nominal frequency	60 Hz	
Voltage limit (Steady state)	Base 345 kV Base 154 kV	0.95~1.05 p.u. 0.90~1.10 p.u.
Voltage limit (Transient state)	Base 345 kV Base 154 kV	0.90~1.05 p.u. 0.90~1.10 p.u.
Branch Overload limit (Steady state)	Branch	100%
Branch Overload limit (Transient)	Line Transformer	120% 100%

Table 3. Specification of BTB VSC HVDC.

Specification	BTB VSC HVDC
Site	1410–1411
Active power capacity	200 × 2 MW
Reactive power capacity	100 × 2 MVAR
MVA rating	224 × 2 MVA
AC terminal voltage	154 kV
DC voltage	100 kV
Configuration	Double monopole
Conversion loss	1.8% of active power
AC voltage setpoint	Depend on the system condition

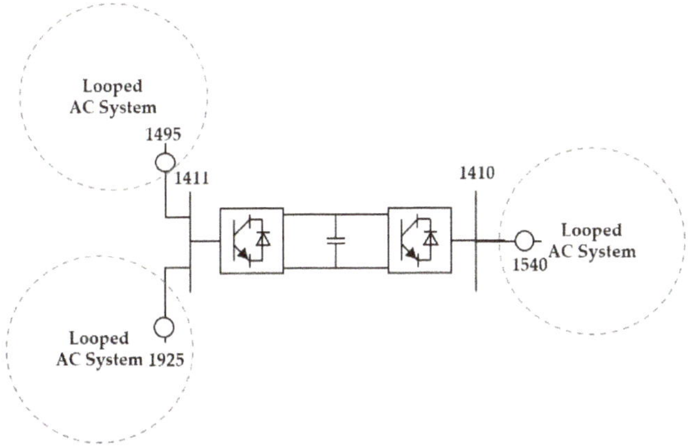

Figure 6. BTB siting in the Korean metropolitan area (1411–1410).

4.2. BTB VSC HVDC Siting for Fault Current Reduction

The site of the BTB VSC HVDC is selected to solve the fault current problem in the Korean metropolitan area and its effect is shown in Table 4. With the BTB VSC HVDC, there is only one bus, bus 1200, at which the fault current is above the 50 kA limit, whereas without the BTB VSC HVDC, there are multiple buses at which fault currents are above the 50 kA limit. The benefit of the BTB VSC HVDC for reducing the fault current is demonstrated.

Table 4. Fault current comparison with/without HVDC [kA].

Bus Number	Base kV	With HVDC	Without HVDC	Bus Number	Base kV	With HVDC	Without HVDC
1200	345	54.8899	55.0129	1810	154	48.2441	53.2228
1400	345	48.2340	50.1957	1811	154	47.7596	52.6257
1410	154	49.7672	65.5842	1865	154	46.2224	51.5447
1411	154	25.7956	65.5842	1870	154	47.0455	52.7775
1475	154	46.1315	51.8724	1895	154	47.1782	53.1212
1490	154	46.2564	52.0078	1955	154	44.5498	50.1641

4.3. STLM Strategy in Normal Condition

Before determining the operating point in normal operation, Equations (7)–(9) were applied to reduce the number of observed branches to be considered in the overload constraint. The active power capacity P_{cap} is 400 MW, weighting factor W is 0.3, and other necessary values were applied using Korean power system data. The monitoring branches determined using Equations (7)–(9) are in Appendix A.

To apply the STLM strategy, the initial conditions must be found. The initial conditions of this case study are as follows.

- Initial operating point $S_0^{(1)}$: (0, 0, 0),
- Independent directions used for initial searching $\epsilon = \{\vec{P}, \vec{Q_{rec}}, \vec{Q_{inv}}\}$,
- Step size $\delta = 1$.

The loss reduction from the operating point found by using the STLM can be seen in Tables 5 and 6. For the case when the HVDC is in operation, we compared the losses for (1) the case where the STLM strategy is applied and (2) the case where the operating condition is applied to the amount of power flow in the existing AC line without applying the STLM. The HVDC operating point in the case of "with HVDC (AC flow)" does not have a special strategy, and the operating point is determined based on the power flow through the AC line when there is no HVDC (when connected to the AC line instead of the separation between buses). More than 400 MW of power flowed to the AC line before the bus separation, and 200 MW × 2 was selected as the operation point according to the HVDC rating limit. When the STLM is applied, the system losses are the least and the technology of the conversion loss of the converter further develops, and the economy caused by the reduction of the loss can be improved. Additionally, at the operating point obtained by applying the STLM strategy, there are no overload lines, and the maximum/minimum voltage of the system is 1.060/0.984 at the 154 kV level, and 1.039/0.995 at the 345 kV level. These values meet the system overload and voltage criteria described in Table 2.

Table 5. HVDC setpoint using the STLM control.

	P (1410 → 1411)	AC Voltage (Rectifier)	AC Voltage (Inverter)
Setpoint	131 × 2 MW	1.048 kV (Supply 95.31 MVAR)	1.039 kV (Supply 3.80 MVAR)

Table 6. Comparison of system losses by operating method.

Losses	With HVDC (STLM Control)	With HVDC (AC Flow)	Without HVDC
AC system losses	1564.38 MW	1578.10 MW	1571.16 MW
DC system losses	4.72 MW	7.20 MW	-
Total losses	1569.10 MW	1585.30 MW	1571.16 MW

4.4. SCLM Control in Transient Condition

An example of the SCLM control for an overload condition that causes N-1 failures in a given system is shown. The transformer overload (over 100%) and the line overload (over 120%) are shown as examples. The simulation proceeds as shown in Figure 7. A fault occurred in 1 s; after six cycles (0.1 s) clear fault, trip line; six cycles later, the HVDC operating point was changed for branch overloads. In this process, fast operation is required to solve the thermal limit, so only active power control is used and the *AC* voltage control setpoint does not change. In addition, the operation point was changed again for loss minimization after 1 s from the failure occurrence.

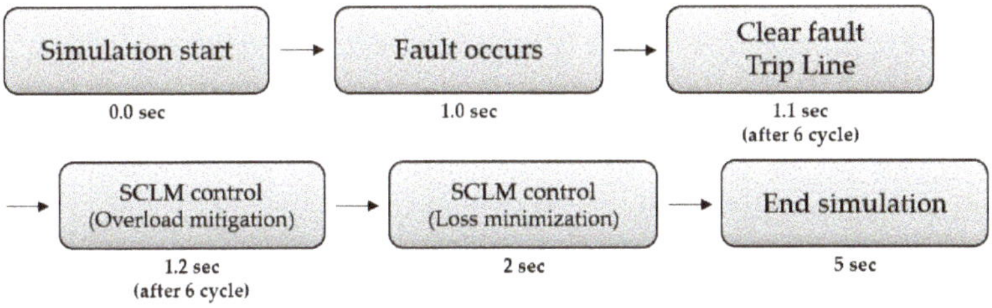

Figure 7. Simulation proceeds for SCLM control.

In order to verify the effect of the SCLM control, this paper compares the results of conventional control, RAS control, and the proposed control. In conventional control, regardless of the disturbance, the active power and *AC* voltage setpoint of the HVDC are constant [33]. Therefore, it may not respond to disturbance, and overload and overvoltage may occur. In this case, the overload condition is maintained, and the transformer is in trouble. In the RAS control, as in the literature reviews, the operating point of the HVDC changes for system stabilization [6–16]. Using the control, the system instability is solved by the active power control, but the AC voltage setpoint is constant without control. Similar to RAS control, the SCLM control solves the overload problem with active power change, and then uses the STLM control to determine the operating point after a contingency. In this case, the operation point is changed first, and in the SCLM control considering the loss minimization, it can be confirmed that the operation point is changed secondarily.

4.4.1. Transformer Overload (100%) Case

When a 1400–1800 2CKT contingency occurs, the overload of the transformer will be 112.51%. At this time, the overload is solved by using SCLM control, and then the operating point for loss minimization is found. The change in the operating point of the HVDC after the SCLM control and the result of the system control are shown in Tables 7 and 8 and Figures 8–11. As in the previous STLM control, after SCLM control, there are no overload lines, and the maximum/minimum voltage of the system is 1.059/0.984 at the 154 kV level, and 1.038/0.933 at the 345 kV level, which also meets the voltage criterion in Table 2.

Table 7. HVDC setpoint change using the SCLM control for the transformer overload.

Control Mode	P (1410 → 1411)	AC Voltage (Rectifier)	AC Voltage (Inverter)
Normal operation (STLM control)	131 × 2 MW	1.048 kV (Supply 95.31 MVAR)	1.039 kV (Supply 3.80 MVAR)
SCLM control	17 × 2 MW	1.019 kV (Supply 82.83 MVAR)	1.048 kV (Supply 72.70 MVAR)

Table 8. Comparison of system losses by operating change (transformer overload case).

	Normal Operation	Conventional Control	RAS Control	SCLM Control
AC system losses	1564.38 MW	1583.92 MW	1587.21 MW	1585.80 MW
DC system losses	4.72 MW	4.72 MW	0.86 MW	0.61 MW
Total losses	1569.10 MW	1587.64 MW	1588.07 MW	1586.41 MW
% Loading	66.02%	110.44% (Security violated)	98.08%	96.41%

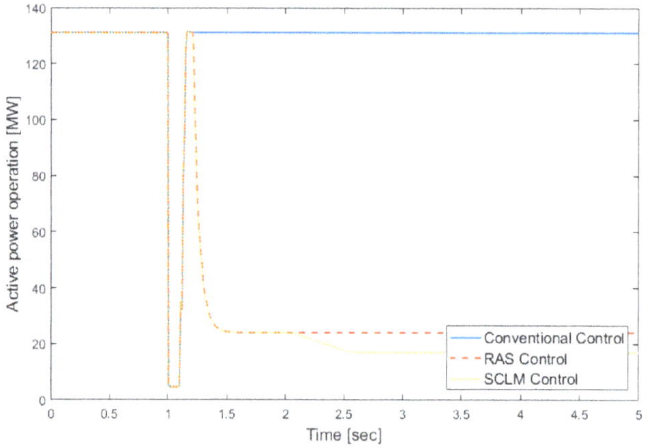

Figure 8. Active power operation change in the transformer overload case.

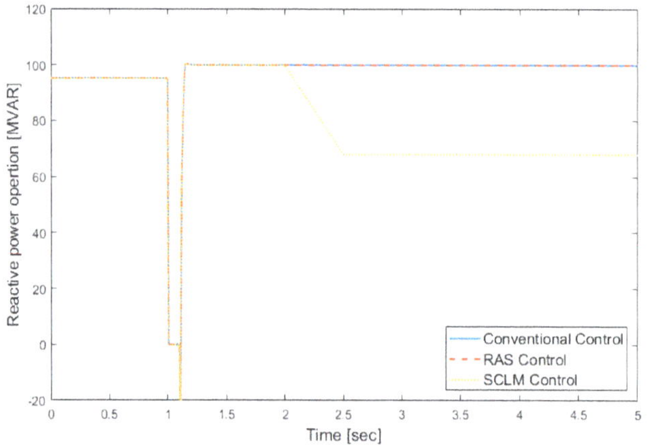

Figure 9. Reactive power operation change in the transformer overload case at the rectifier end.

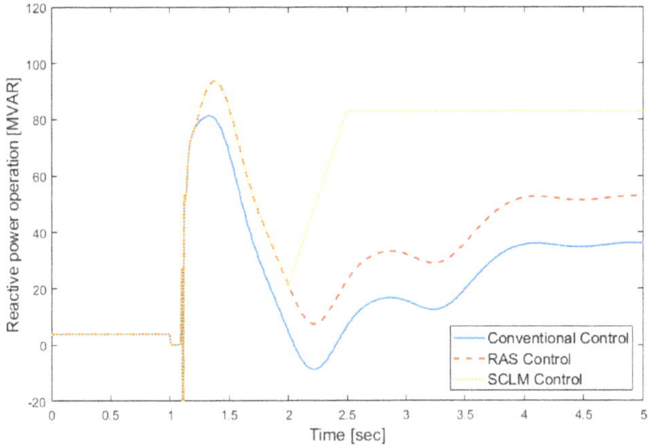

Figure 10. Reactive power operation change in the transformer overload case at the inverter end.

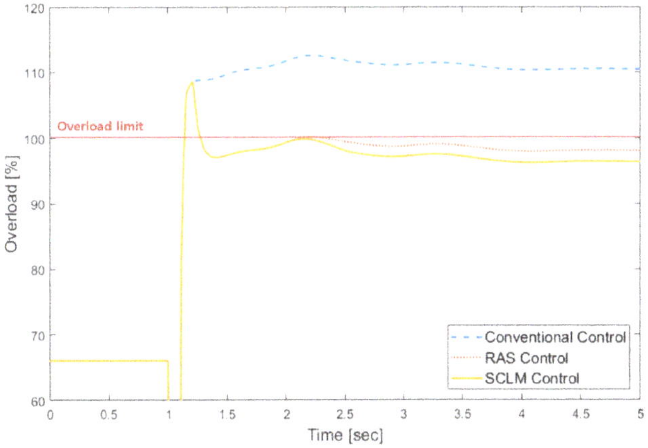

Figure 11. Comparison of HVDC control for overload mitigation in the transformer overload case.

4.4.2. Line Overload (120%) Case

When a 1395–1415 1CKT contingency occurs, the overload of the remaining line will be 123.42%. Like the previous case, the overload is solved through the SCLM control, and then the operating point for minimum loss is determined. The change in the operating point of the HVDC after the SCLM control and the result of the system control are shown in Tables 9 and 10 and Figures 12–15. Additionally, in this case, there are no overload lines, and the maximum/minimum voltage of the system is 1.059/0.984 at the 154 kV level, and 1.039/0.994 at the 345 kV level, which also meets the voltage criterion in Table 2.

Table 9. HVDC setpoint change using the SCLM control for line overload.

Control Mode	P (1410→1411)	AC Voltage (Rectifier)	AC Voltage (Inverter)
Normal operation (STLM control)	131 × 2 MW	1.048 kV (Supply 95.31 MVAR)	1.039 kV (Supply 3.80 MVAR)
SCLM control	184 × 2 MW	1.035 kV (Absorb 2.34 MVAR)	1.042 kV (Supply 79.86 MVAR)

Table 10. Comparison of system losses by operating change (line overload case).

	Normal Operation	Conventional Control	RAS Control	SCLM Control
AC system losses	1564.38 MW	1575.28 MW	1573.28 MW	1568.02 MW
DC system losses	4.72 MW	4.72 MW	7.20 MW	6.62 MW
Total losses	1569.10 MW	1580.00 MW	1580.48 MW	1574.64 MW
% Loading	62.98%	122.27% (Security violated)	116.17%	119.08%

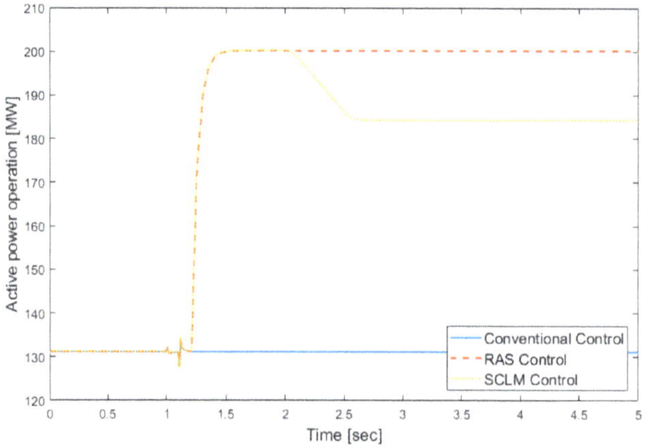

Figure 12. Active power operation change in the line overload case.

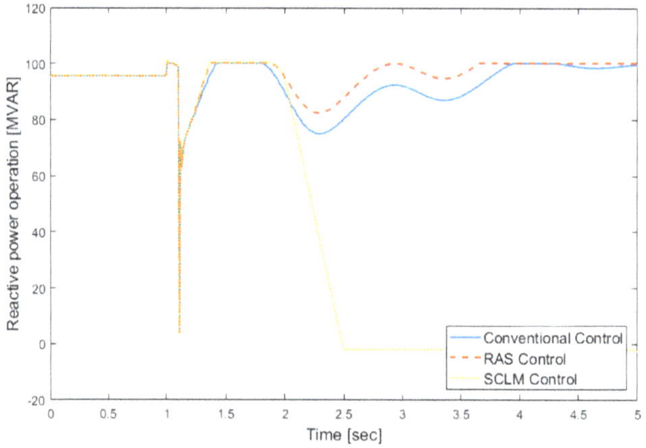

Figure 13. Reactive power operation change in the line overload case at the rectifier end.

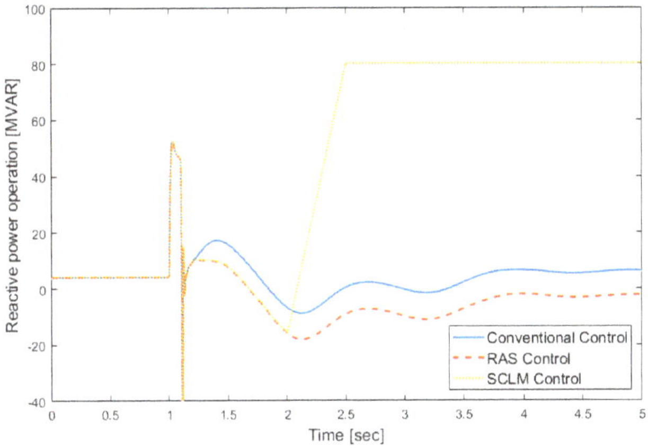

Figure 14. Reactive power operation change in the line overload case at the inverter end.

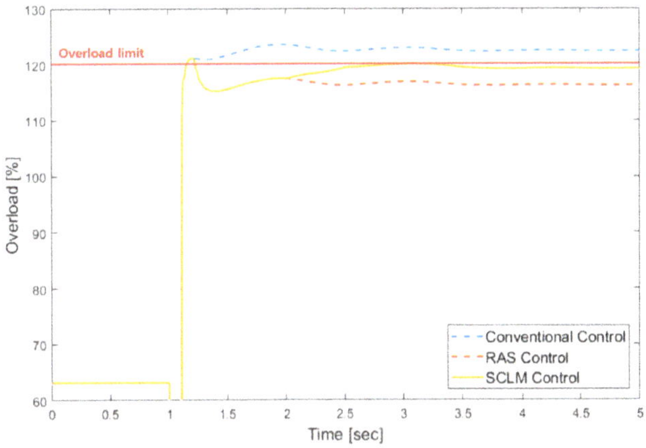

Figure 15. Comparison of HVDC control for overload mitigation in the line overload case.

4.4.3. Comparison of Simulation Results

As shown in Figures 8 and 12, in the conventional control, the HVDC system is controlled to sustain the constant power output, but when the RAS or the SCLM control is applied, the active power setpoint is changed to mitigate the overload. The reason why the active power is changed again after 2 s in the SCLM control is that the operating point is changed to minimize the system total loss within the operating range where the overload is mitigated. Additionally, Figure 9, Figure 10, Figure 13, and Figure 14 show the reactive power change at both ends of HVDC. In the conventional control and RAS control, the converters control the reactive power output to sustain the AC voltage setpoint in the pre-contingency condition, and the reactive power output fluctuated according to the system conditions. On the other hand, in the SCLM control, the reactive power output of HVDC for minimizing the system total loss is calculated and changed after 2 s. Accordingly, the total loss is 1.66 and 5.84 MW lower, respectively, compared to the RAS results as expressed in Tables 8 and 10.

Comparisons of overload changes for each case can be seen in Figures 11 and 15. In terms of the overload criteria, the load of the SCLM control is lower than that of RAS control in Figure 11, whereas the load of RAS control is lower than that of the SCLM control in Figure 15. In the SCLM control, since the operating point is determined to minimize the loss within the operating range for relieving the overload, the line load can be high, but it does not exceed the overload limit, so it is not a problem in terms of reliability.

5. Conclusions

This paper proposes a methodology for determining the operating point of embedded HVDC considering system losses and reliability criterion. In order to verify the effectiveness of the proposed control, we applied the methodology to the Korean power system, as a result, we proved that embedded BTB VSC HVDC can provide more efficient grid service for the AC system as follows.

- In normal operation, the system loss is reduced by about 16 MW in the STLM control. This is 1% of the total loss, but an economic benefit can be gained during the lifetime of HVDC. Although we did not perform a detailed economic assessment, it is possible to obtain a loss reduction effect of about 85 million dollars a year considering energy charge and demand charge.
- There are about 180 branches in the zone with HVDC. Using sensitivity analysis, the number of monitoring branches can be reduced to 14 lines and three transformers.
- If a disturbance occurs in the *AC* system and there is no appropriate HVDC control accordingly, HVDC may adversely affect the *AC* system. Using the SCLM control, not only overload and overvoltage problems are solved, but also the system loss is reduced, which is advantageous in economy and reliability.

The proposed scheme can be operated well in a real system if a large-enough communication system like Supervisory control and data acquisition (SCADA) is applied. It is possible to reduce the communication burden by reducing the number of monitoring lines by precomputing the line that is greatly influenced by the HVDC control. Future research will need to be done on the overall embedded VSC HVDC operation strategy, not limited to the active/reactive power control, but considering all the other grid services it can provide.

Author Contributions: J.L. and M.Y. contributed equally to this work. The main algorithm and control scheme were proposed by J.L. and M.Y. The experiment results were collected by S.H., S.J. (Soseul Jeong) and S.J. (Seungmin Jung). The simulation results were analyzed by J.L., M.Y., G.J., J.L. and M.Y. wrote the paper.

Funding: This research received no external funding.

Acknowledgments: This work was supported by Korea Electric Power Corporation grant (R17XA05-4) and "Human Resources Program in Energy Technology" of the Korea Institute of Energy Technology Evaluation and Planning (KETEP), granted financial resource from the Ministry of Trade, Industry & Energy, Republic of Korea. (No.20174030201540).

Conflicts of Interest: The authors declare no conflict of interest.

Appendix A

Table A1. Monitoring branches for HVDC operation.

1310–1340	1340–1825	1360–1945	1361–1945	1411–1495	1411–1925
1495–1690	1690–1945	1825–1890	1865–1870	1890–1935	1925–1931
1931–1940	1935–1940	1360 M.Tr	1361 M.Tr	1410 M.Tr	

References

1. Van Hertem, D.; Ghandhari, M. Multi-terminal VSC HVDC for the European supergrid: Obstacles. *Renew. Sustain. Energy Rev.* **2010**, *14*, 3156–3163. [CrossRef]
2. Van Hertem, D.; Gomis-Bellmunt, O.; Liang, J. *HVDC Grids: For Offshore and Supergrid of the Future*; John Wiley & Sons: Hoboken, NJ, USA, 2016; ISBN 9781118859155.
3. Peake, O. The history of high voltage direct current transmission. *Aust. J. Multi Discip. Eng.* **2010**, *8*, 47–55. [CrossRef]

4. Renner, R.H.; Van Hertem, D. Ancillary services in electric power systems with HVDC grids. *IET Gener. Transm. Distrib.* **2015**, *9*, 1179–1185. [CrossRef]
5. Henry, S.; Despouys, O.; Adapa, R.; Barthold, C.; Bell, K.; Binard, J.-L.; Edris, A.; Egrot, P.; Hung, W.; Irwin, G.; et al. Influence of embedded HVDC transmission on system security and AC network performance. *Electra* **2013**, *267*, 79–86.
6. Johnson, R.K.; Klemm, N.S.; De Laneuville, H.; Koetschau, S.G.; Wild, G. Power modulation of Sidney HVDC scheme. I. RAS control concept, realization and field tests. *IEEE Trans. Power Deliv.* **1989**, *4*, 2145–2152. [CrossRef]
7. Johnson, R.K.; Klemm, N.S.; Schiling, K.H.; Thumm, G. Power modulation of Sidney HVDC scheme. II. Computer simulation. *IEEE Trans. Power Deliv.* **1989**, *4*, 2153–2161. [CrossRef]
8. Lee, R.L.; Melvold, D.J.; Szumlas, D.J.; Le, L.M.; Finley, A.T.; Martin, D.E.; Wong, W.K.; Dickmander, D.L. Potential DC system support to enhance AC system performance in the Western United States. *IEEE Trans. Power Syst.* **1993**, *8*, 264–274. [CrossRef]
9. Cholley, P.; Crossley, P.; Van Acker, V.; Van Cutsem, T.; Fu, W.; Soto Idia Òez, J.; Ilar, F.; Karlsson, D.; Kojima, Y.; McCalley, J.; et al. System protection schemes in power networks. In Proceedings of the 2001 Seventh International Conference on Developments in Power System Protection (IEE), Amsterdam, The Netherlands, 2001; pp. 450–453.
10. Luo, J.; Gong, Y.; Li, H.; Yan, Y. Online emergency control and corrective control coordination strategy for UHVDC blocking faults. In Proceedings of the 2018 IEEE International Conference on Electronics Technology (ICET 2018), Chendu, China, 23–27 May 2018; pp. 5–10.
11. Jiang, M.; Yu, H. Alberta's experience of coordinating HVDC operation with under-voltage remedial action scheme. In Proceedings of the 2016 IEEE Power and Energy Society General Meeting (PESGM 2016), Boston, MA, USA, 17–21 July 2016; pp. 1–4.
12. Bhuiyan, M.R.; Taylor, A.E.; Kezman, N.C. Designing the hardware and software of the Pacific HVDC intertie remedial action scheme using a programmable controller. In Proceedings of the 1991 IEEE Power Engineering Society Transmission and Distribution Conference, Dallas, TX, USA, 22–27 September 1991; pp. 790–796.
13. Dolezilek, D. Case study examples of interoperable ethernet communications within distribution, transmission, and wide-area control systems. In Proceedings of the 2010 IEEE International Conference on Communications Workshops (ICC 2010), Cape Town, South Africa, 23–27 May 2010; pp. 1–7.
14. Bahrman, M.; Bjorklund, P.-E. The new black start: System restoration with help from voltage-sourced converters. *IEEE Power Energy Mag.* **2014**, *12*, 44–53. [CrossRef]
15. Yoon, M.; Yoon, Y.-T.; Jang, G. A study on maximum wind power penetration limit in island power system considering high-voltage direct current interconnections. *Energies* **2015**, *8*, 14244–14259. [CrossRef]
16. Miao, Y.; Cheng, H. An optimal reactive power control strategy for UHVAC/DC hybrid system in east china grid. *IEEE Trans. Smart Grid* **2016**, *7*, 392–399. [CrossRef]
17. Moslehi, K. Electricity market challenges—Efficient utilization of transmission. In Proceedings of the 2015 IEEE Smart Grids Seminar, Cuernavaca, Morelos, Mexico, 23–27 March 2015.
18. Cao, J.; Du, W.; Wang, H.F.; Bu, S.Q. Minimization of transmission loss in meshed AC/DC grids with VSC-MTDC networks. *IEEE Trans. Power Syst.* **2013**, *28*, 3047–3055. [CrossRef]
19. Han, M.; Xu, D.; Wan, L. Hierarchical optimal power flow control for loss minimization in hybrid multi-terminal HVDC transmission system. *CSEE J. Power Energy Syst.* **2016**, *2*, 40–46. [CrossRef]
20. QI, Q.; Long, C.; Wu, J.; Yu, J. Impacts of a medium voltage direct current link on the performance of electrical distribution networks. *Appl. Energy* **2018**, *230*, 175–188. [CrossRef]
21. QI, Q.; Wu, J.; Long, C. Multi-objective operation optimization of an electrical distribution network with soft open point. *Appl. Energy* **2017**, *208*, 734–744. [CrossRef]
22. Vovos, P.N.; Song, H.; Cho, K.-V.; Kim, T.-S. A network reconfiguration algorithm for the reduction of expected fault currents. In Proceedings of the 2013 IEEE Power and Energy Society General Meeting (PES 2013), Vancouver, BC, Canada, 21–25 July 2013; pp. 1–5.
23. Lee, J.-H.; Yoon, M.; Jung, S.; Jang, G. System reliability enhancement in a metropolitan area using HVDC technology. *J. Int. Counc. Electr. Eng.* **2015**, *5*, 1–5. [CrossRef]

24. Tongsiri, S.; Hoonchareon, N. Fault current limitation in metropolitan power system using HVDC link. In Proceedings of the 8th IEEE Electrical Engineering/Electronics, Computer, Telecommunications and Information Technology (ECTI-CON 2011), Khon Kaen, Thailand, 17–20 May 2011; pp. 840–844.
25. Pang, H.; Tang, G.; He, Z. Evaluation of losses in VSC-HVDC transmission system. In Proceedings of the 2008 IEEE Power and Energy Society General Meeting-Conversion and Delivery of Electrical Energy in the 21st Century, Pittsburgh, PA, USA, 20–24 July 2008; pp. 1–6.
26. Sellick, R.L.; Åkerberg, M. Comparison of HVDC light (VSC) and HVDC classic (LCC) site aspects, for a 500 MW 400 kV HVDC transmission scheme. In Proceedings of the 10th IET International Conference on AC and DC Power Transmission (ACDC 2012), Birmingham, UK, 4–5 December 2012.
27. Fradi, A.; Brignone, S.; Wollenberg, B.E. Calculation of energy transaction allocation factors. *IEEE Trans. Power Syst.* **2001**, *16*, 266–272. [CrossRef]
28. Song, C.S.; Park, C.H.; Yoon, M.; Jang, G. Implementation of PTDFs and LODFs for power system security. *J. Int. Counc. Electr. Eng.* **2011**, *1*, 49–53. [CrossRef]
29. Rahmani, M.; Kargarian, A.; Hug, G. Comprehensive power transfer distribution factor model for large-scale transmission expansion planning. *IET Gener. Transm. Distrib.* **2016**, *10*, 2981–2989. [CrossRef]
30. Powell, M.J.D. An efficient method for finding the minimum of a function of several variables without calculating derivatives. *Comput. J.* **1964**, *7*, 155–162. [CrossRef]
31. Lazarou, S.; Vita, V.; Ekonomou, L. Application of powell's optimisation method for the optimal number of wind turbines in a wind farm. *IET Sci. Meas. Technol.* **2011**, *5*, 77–80. [CrossRef]
32. Cao, W.; Wu, J.; Jenkins, N. Feeder load balancing in MV distribution networks using soft normally-open points. In Proceedings of the 2014 IEEE PES Innovative Smart Grid Technologies Conference Europe (ISGT-Europe 2014), Istanbul, Turkey, 12–15 October 2014; pp. 1–6.
33. Asplund, G.; Eriksspon, K.; Jiang, H.; Lindberg, J.; Pålsson, R.; Swensson, K. DC transmission based on voltage source converters. In *CIGRE SC14 Colloquium, South Africa*; ABB Power Systems AB: Vasteras, Sweden, 1997; pp. 1–7.

applied sciences

MDPI

Article

A Novel Overcurrent Suppression Strategy during Reclosing Process of MMC-HVDC

Bin Jiang * and Yanfeng Gong

Department of Electrical and Electronic Engineering, North China Electric Power University, Beijing 102206, China; yanfeng.gong@ncepu.edu.cn

* Correspondence: jiangbin911@foxmail.com; Tel.: +86-10-6177-1571

Received: 28 March 2019; Accepted: 25 April 2019; Published: 26 April 2019

Abstract: A modular multilevel converter based high-voltage DC (MMC-HVDC) system has been the most promising topology for HVDC. A reclosing scheme is usually configured because temporary faults often occur on transmission lines especially when overhead lines are used, which often brings about an overcurrent problem. In this paper, a new fault current limiter (FCL) based on reclosing current limiting resistance (RCLR) is proposed to solve the overcurrent problem during the reclosing process. Firstly, a mesh current method (MCM) based short-circuit current calculation method is newly proposed to solve the fault current calculation of a loop MMC-HVDC grid. Then the method to calculate the RCLR is proposed based on the arm current to limit the arm currents to a specified value during the reclosing process. Finally, a three-terminal loop MMC-HVDC test grid is constructed in the widely used electromagnetic transient simulation software PSCAD/EMTDC and the simulations prove the effectiveness of the proposed strategy.

Keywords: modular multilevel converter (MMC); reclosing process; fault current limiter (FCL); short-circuit current calculation; reclosing current limiting resistance (RCLR)

1. Introduction

The modular multilevel converter (MMC), a new type of voltage source converter (VSC), has many advantages over the traditional two or three level topology, such as better harmonic performance, low switching frequency, etc. This advanced technology is envisioned as the most promising option for the integration of renewable energy sources and has been widely investigated in recent years [1–3]. In China, an MMC-HVDC grid usually uses overhead lines because large-scale renewable energy bases are generally far from load centers, as it makes the fault ride-through of MMC-HVDC grid become more thorny [4].

Currently, the half-bridge sub-module (HBSM) is the main topology of a practical MMC-HVDC grid because of better economy [5]. When a short-circuit fault occurs on the DC line, very large fault currents will be caused in the DC lines and converters [6]. Since temporary faults often occur on overhead lines, MMC-HVDC needs reclosing to improve the reliability and continuity of the power supply [7]. Consequently, the vulnerable power electronic elements in the converter will be exposed to overcurrent again when the system attempts to reclose a permanent fault.

The suppression of overcurrent mainly depends on an effective fault current limiter (FCL). Reference [8] limited the overcurrent in HVDC lines by a current limiting inductor (CLI) and studied the impact of different sizes on current rise. Reference [9,10] studied the performance of a superconducting fault current limiter (SFCL) and concluded that the appropriated SFCL can reduce the fault current peak, the size of the CLI and also the fault identification time. Reference [11] presented a solid state FCL in a VSC-HVDC system with MMC converters, but the detailed structure and the configuration method were not clear. Reference [12,13] both proposed a resistive DC fault current limiter. The limiter was installed in each arm or at the outlet of the converter to increase the damping performance of the DC fault current loop and effectively limit the dc-side current, and then interrupt the fault circuit by a breaker. Reference [14] proposed a RL-FCL which is composed of a reactor, a resistor, a reversed diode and an IGBT connected in parallel to limit the overcurrent and the overvoltage. The aforementioned references mainly focused on the overcurrent limiting when the fault occurred but did not conduct further research on the post-fault and the reclosing process.

There has been little research on reclosing process, especially for MMC-HVDC. Reference [15] used two thyristors in parallel with the sub-modules to realize fault clearance and automatic recovery, but the converter must be blocked for a long time when reclosing a permanent fault. Reference [16] used series braking resistance (SBR) cooperating with a high-temperature superconducting FCL to reduce the fault currents during reclosing, but it did not illustrate how to select the parameter of the SBR. A soft reclosing model (SRM) is proposed in Reference [17] to limit the reclosing over-current of VSC-MTDC. However, the current calculation in this reference is made on the assumption that the voltage discharge voltage remains constant and neglects the restriction of the safety of the equipment. Overcurrent suppression during the reclosing process of MMC-HVDC needs further research.

This paper focuses on the suppression of overcurrent for overhead lines based MMC-HVDC grid during the reclosing process. First, mesh current method (MCM) based short-circuit current calculation is introduced to a loop MMC-HVDC grid. Then a novel FCL, which consists of a CLI, a reclosing current limiting resistance (RCLR) and a bypass switch, is proposed. Compared with traditional FCL, the proposed circuit can limit the maximum arm currents to a specified value during the reclosing process, and the calculation method of RCLR is also given in details.

2. Fault Current Calculation of a Loop MMC-HVDC Grid

2.1. Characteristics of a Pole-to-Pole Fault

A common three-phase MMC is shown in Figure 1. It is composed of six arms and each arm consists of N sub-modules (SMs) connected in series and an arm inductor L_0. Each SM includes two insulated gate bipolar transistors (IGBTs), two diodes and a DC capacitor C_0. If the SM is 'ON' state, the output of SM is the dc capacitor voltage. When the SM is 'OFF' state, the output of SM is zero.

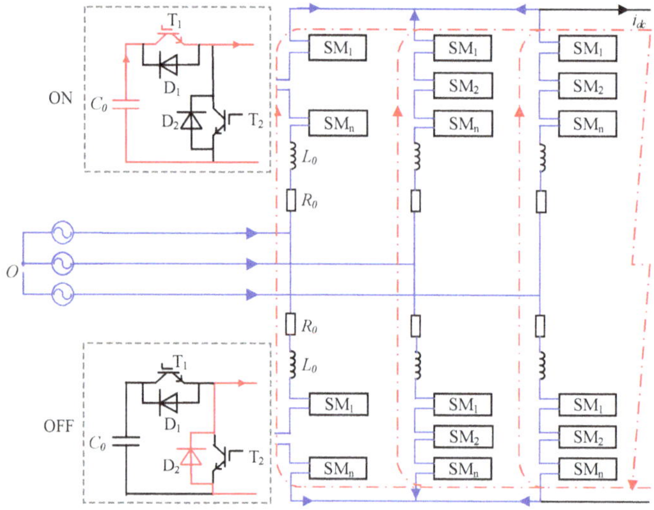

Figure 1. Fault current path of a pole-to-pole fault before converter blocking.

The pole-to-pole fault is the most serious fault in the DC side of MMC-HVDC. The fault process can be divided into two stages according to whether the converter is blocked [18].

Before the converter is blocked, the arm current is shown in Figure 1. During this stage, all the capacitors of SMs in 'ON' state will discharge through T_1 immediately (the red line), and thus the currents in the arms and DC lines increase extremely fast. For the AC system, the fault is equivalent to a three-phase short-circuit. So, the AC system also injects short-circuit currents (the blue line) during this stage, whereas they are not capable of flowing into the DC lines, because their contributions sum to zero in the converter. The arm current is a superposition of the discharging current and three-phase short-circuit current. Due to the sorting algorithm within the converter, there are always N SMs being switched in thus all the SMs will discharge during the fault. Given to the high control frequency, the SMs in the upper and the lower arm can be regarded as discharging at the same time. Therefore, the equivalent circuit of the converter during this stage is displayed as Figure 2.

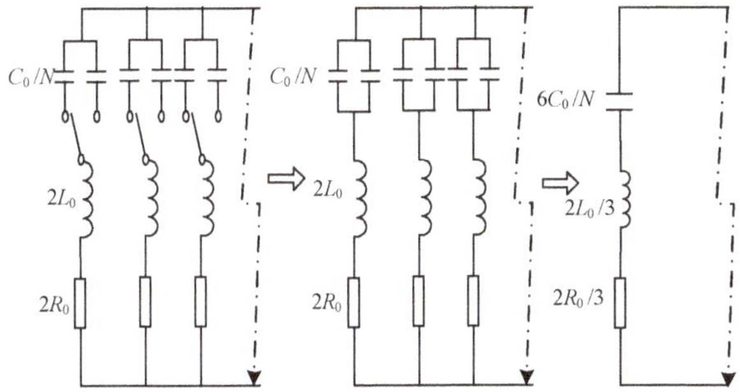

Figure 2. Equivalent circuit of the converter under a pole-to-pole fault before converter blocking.

If the arm current in any arm exceeds the threshold of overcurrent protection of SM, the converter will be blocked immediately. The fault process after blocking can be further divided into two sub-stages. The first one is inductor free-wheeling circuit as is shown in Figure 3a. The fault currents are maintained by the arm inductor. At this stage, each arm still carries a third of the DC line current, and no AC currents are injected in DC line.

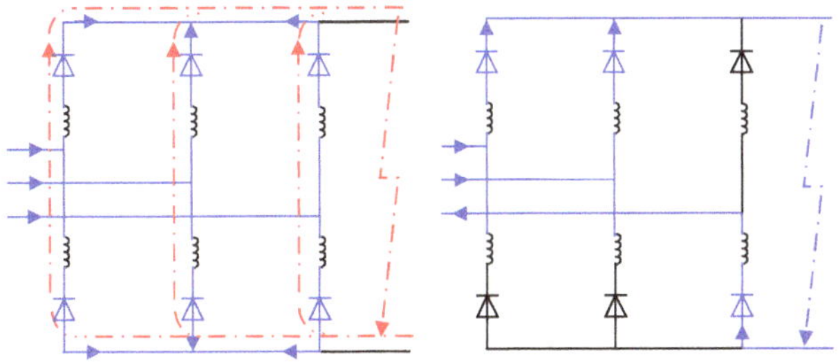

Figure 3. The fault response after converter blocking: (**a**) free-wheeling; (**b**) uncontrolled rectifier.

The stage of inductor free-wheeling will end when the inductor free-wheeling current decrease to zero. Then the converter starts to act as an uncontrolled rectifier as Figure 3b. At this stage, the fault currents in the DC lines are fed by the AC system. Generally, the protection time in MMC-HVDC grid is within 5 ms after the fault occurrence [19], so this stage of the uncontrolled rectifier seldom occurs in practical system.

2.2. Short-Circuit Current Calculation Method

The short-circuit current calculation of a loop MMC-HVDC is different from that of a two-terminal MMC-HVDC or a star join multi terminal DC grid as the fault current of the latter can be calculated separately. For a loop MMC-HVDC, as displayed in Figure 4, all the converters inject fault current through different lines to the fault spot, and the currents from different converters cause coupling problem.

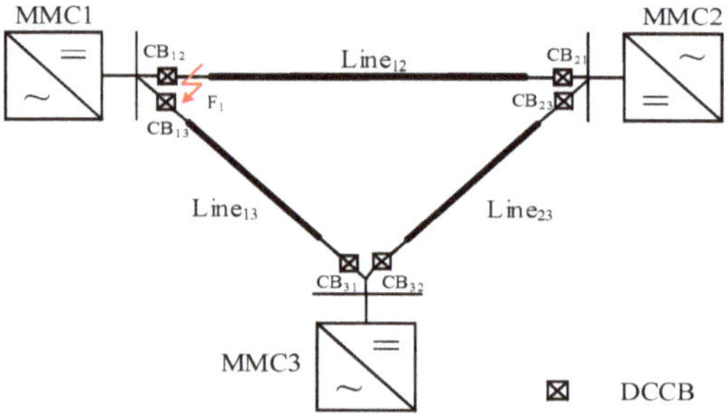

Figure 4. The diagram of the three-terminal loop modular multilevel converter based high-voltage DC (MMC-HVDC) grid.

To handle the current coupling problem, the mesh current method (MCM) is first introduced to a short-circuit current calculation of MMC-HVDC grid before the converters are blocking. For example, when there is a pole-to-pole fault at the DC outlet of MMC1, the whole grid can be divided into four meshes. The equivalent circuit is illustrated in Figure 5. L_i, C_i and R_i represent the equivalent circuit of the converter, r_{ij} and l_{ij} represent the DC overhead lines and L_{ij} is CLI. According to the four meshes, a state equation can be established as Equations (1) and (2). Then the current in different lines can be calculated.

$$\begin{bmatrix} 1 & 0 & 0 \\ 1 & 0 & -1 \\ 0 & 1 & -1 \\ 0 & 1 & 0 \end{bmatrix} \begin{bmatrix} u_{c1} \\ u_{c2} \\ u_{c3} \end{bmatrix} = \begin{bmatrix} R_1 & R_1 & 0 & 0 \\ R_1 & 2r_{13}+R_1+R_3 & R_3 & 0 \\ 0 & R_3 & 2r_{23}+R_2+R_3 & R_2 \\ 0 & 0 & R_2 & 2r_{12}+R_2 \end{bmatrix} \begin{bmatrix} i_{14} \\ i_{13} \\ i_{23} \\ i_{24} \end{bmatrix} + \begin{bmatrix} 2L_{12}+L_1 & L_1 & 0 & 0 \\ L_1 & 2(L_{13}+L_{31}+l_{13})+L_1+L_3 & L_3 & 0 \\ 0 & L_3 & 2(L_{23}+L_{32}+l_{23})+L_2+L_3 & L_2 \\ 0 & 0 & L_2 & 2(L_{21}+L_{13})+L_2 \end{bmatrix} \begin{bmatrix} \dot{i}_{14} \\ \dot{i}_{13} \\ \dot{i}_{23} \\ \dot{i}_{24} \end{bmatrix} \quad (1)$$

$$\begin{bmatrix} \dot{u}_{c1} \\ \dot{u}_{c2} \\ \dot{u}_{c3} \end{bmatrix} = \begin{bmatrix} 1/C_1 & 0 & 0 \\ 0 & 1/C_2 & 0 \\ 0 & 0 & 1/C_3 \end{bmatrix} \begin{bmatrix} 1 & 1 & 0 & 0 \\ 0 & 0 & 1 & 1 \\ 0 & -1 & -1 & 0 \end{bmatrix} \begin{bmatrix} i_{14} \\ i_{13} \\ i_{23} \\ i_{24} \end{bmatrix} \quad (2)$$

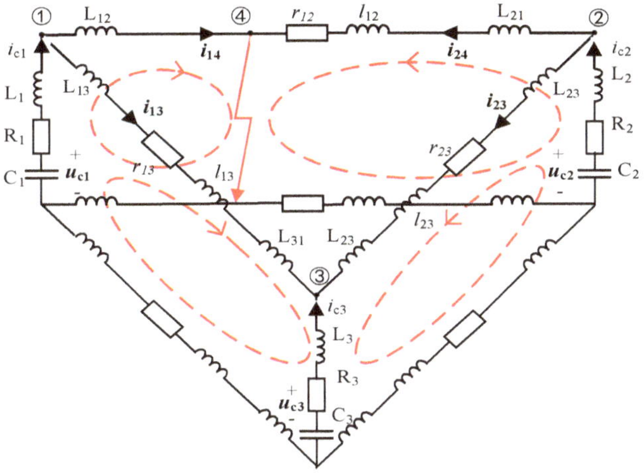

Figure 5. The equivalent circuit of SMs discharging under pole-to-pole fault.

3. Operation Principle of the Reclosing Current Limiting Resistance based Fault Current Limiter

3.1. The Operation Principle of the Fault Current Limiter

The CLI is an effective and mature device to limit fault currents in the MMC-HVDC grid, but it can't further reduce the overcurrent stresses on the power electronic elements in the converter in the reclosing process. Therefore, an additional circuit, which is composed of an RCLR and a parallel switch, is proposed based on the traditional CLI, as shown in the dotted line box of Figure 6.

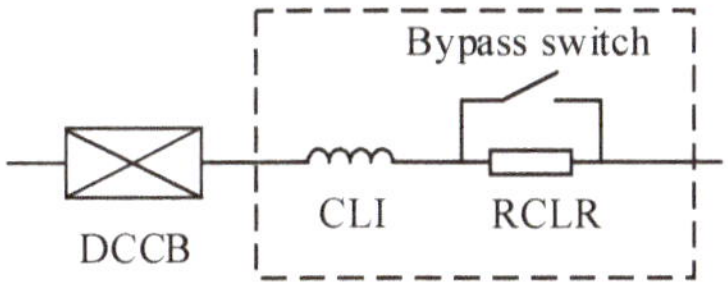

Figure 6. The structure of the proposed current fault limiter.

Similar to an AC system, the fault ride-through of an overhead lines based MMC-HVDC grid needs two or three reclosing attempts to determine whether the fault is temporary. Figure 7 shows the sequence of fault ride-through when reclosing a permanent fault. The fault starts at the instant of t_0, and the DCCB on the faulty line will be tripped at t_{r1} after the protection identification time of Δt_c. During the deionization time of Δt_d, the bypass switch is opened, and thus the RCLR is inserted into the fault loop. The DCCB is reclosed at t_{p1} and tripped again at t_{r2} after detecting there is still a fault. By using the RCLR, the overcurrent stresses on powering electronic elements can be reduced significantly, and the converter can keep continuous operation. The same reclosing process will be done at t_{p2} and t_{r2}. At last, the DCCB will never be reclosed after the identification of a permanent fault. The bypass switch is opened at t_{R2} and waits for the next operation command.

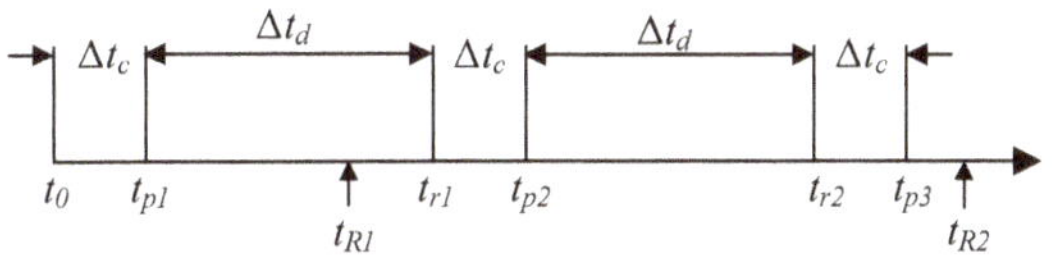

Figure 7. The sequence of fault ride-through when reclosing a permanent fault.

3.2. The Calculation Method of RCLR

There are two main electric elements in the proposed FCL. The main function of CLI is to limit the rise rate of fault currents and ensure the reliable operation of the DCCB, while that of the RCLR is to reduce the overcurrent in the converter during reclosing. Extensive papers have studied the calculation method of the CLI in MMC-HVDC grid [8,20–22], so it will not be discussed in this paper. The main contribution of this paper is to propose a novel calculation method of RCLR, which can limit the arm currents to a specified value. According to the aforementioned analysis, the constraint of the RCLR is denoted by Equation (3). $I_{\max}$ represents the maximum value of arm current during reclosing, I_{th} represents the threshold of SMs overcurrent protection, and k is ratio coefficient which denotes the limiting level of fault currents.

$$I_{\max} \leq k \times I_{th} \tag{3}$$

By using the RCLR, only SMs discharging will appear during reclosing, whereas it is difficult to describe the accurate relations between arm currents and RCLR through mathematical equations. In order to solve this problem, a method of estimating the maximum arm current is proposed. The FCL reduces the discharging speed of the SMs significantly during reclosing, so the AC output can be considered still under control within the several milliseconds after fault. If the circulating currents in the arm are neglected, the maximum current in the arm is approximately equal to Equation (4)

$$I_{arm} = \frac{i_{ac}}{2} + \frac{1}{3}\left|i_{dc}(\Delta t_c)\right| \tag{4}$$

where i_{ac} is the amplitude of phase current in the AC side of converter, i_{dc} represents the current measured at positive pole of the converter shown in Figure 1, and the Δt_c is protection time from reclosing time to the trip time of the DCCB. By substituting (4) into (3), the following inequality can be obtained as Equation (5). Define the value satisfying the equation condition of (5) as the critical resistance R_0.

$$\frac{i_{ac}}{2} + \frac{1}{3}\left|i_{dc}(\Delta t_c)\right| \leq k \times I_{th} \tag{5}$$

4. Simulation

4.1. The Test System

A three-terminal MMC-HVDC grid, shown in Figure 4, is modeled based on PSCAD/EMTDC simulation platform. The system adopts the symmetric monopole configuration with a rated DC voltage of ±20 kV. All the converters are connected by bipolar transmission lines which make use of a frequency-dependent distributed parameter model, and a hybrid DCCB and a FCL are placed at the terminal of each line. MMC1 and MMC3 control the active power, while MMC2 controls the DC voltage. The detailed parameters of the test system are shown in Table 1. Other simulation parameters are set as follows: the protection time Δt_c is 5 ms; the deionization time Δt_d is 150 ms; the threshold of SMs overcurrent protection I_{th} is 2 kA; the CLI on each line is 15 mH. Moreover, the converter will be blocked after the fault current declines to 10 A.

Table 1. The parameters of the test system.

Quantity	Value
AC line voltage	10 kV
Fundamental frequency	50 Hz
Number of SMs per arm	20
SM capacitor	6000 μF
Arm reactor	15 mH
Arm resistance	0.1 Ω
Transformer leakage reactance	0.1

4.2. The Validation of the FCL

For the test system, a pole-to-pole fault F_1 is applied at the DC outlet of MMC1. The overcurrent in MMC1 is the most serious, and thus it will be taken as the example for simulation. When the RCLR R_{12} is configured, the equivalent circuit in Figure 4 is modified as Figure 8 and Equation (1) is modified as Equation (6). The upper and lower limitation of the RCLR are set to 0 and 15 Ω, respectively, and the calculation step is 0.1 Ω.

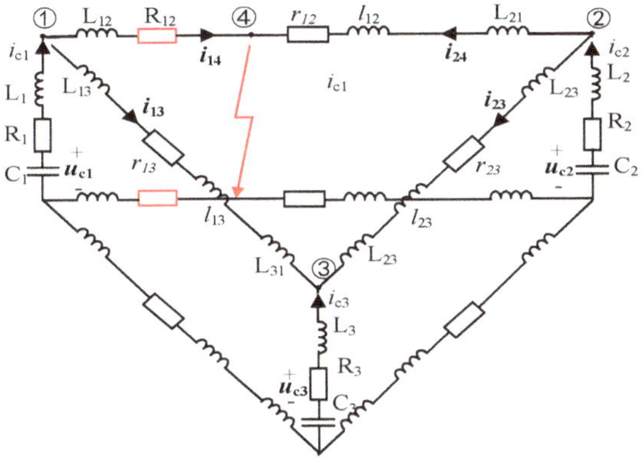

Figure 8. The equivalent circuit of SMs discharging with reclosing current limiting resistance (RCLR).

$$\begin{bmatrix} 1 & 0 & 0 \\ 1 & 0 & -1 \\ 0 & 1 & -1 \\ 0 & 1 & 0 \end{bmatrix} \begin{bmatrix} u_{c1} \\ u_{c2} \\ u_{c3} \end{bmatrix} = \begin{bmatrix} 2R_{12}+R_1 & R_1 & 0 & 0 \\ R_1 & 2r_{13}+R_1+R_3 & R_3 & 0 \\ 0 & R_3 & 2r_{23}+R_2+R_3 & R_2 \\ 0 & 0 & R_2 & 2r_{12}+R_2 \end{bmatrix} \begin{bmatrix} i_{14} \\ i_{13} \\ i_{23} \\ i_{24} \end{bmatrix} + \begin{bmatrix} 2L_{12}+L_1 & L_1 & 0 & 0 \\ L_1 & 2(L_{13}+L_{31}+l_{13})+L_1+L_3 & L_3 & 0 \\ 0 & L_3 & 2(L_{23}+L_{32}+l_{23})+L_2+L_3 & L_2 \\ 0 & 0 & L_2 & 2(L_{21}+L_{13})+L_2 \end{bmatrix} \begin{bmatrix} \dot{i}_{14} \\ \dot{i}_{13} \\ \dot{i}_{23} \\ \dot{i}_{24} \end{bmatrix} \quad (6)$$

By solving (2) and (6), the relation between the maximum arm current and the critical resistance R_0 of FCL can be obtained, as shown in Figure 9.

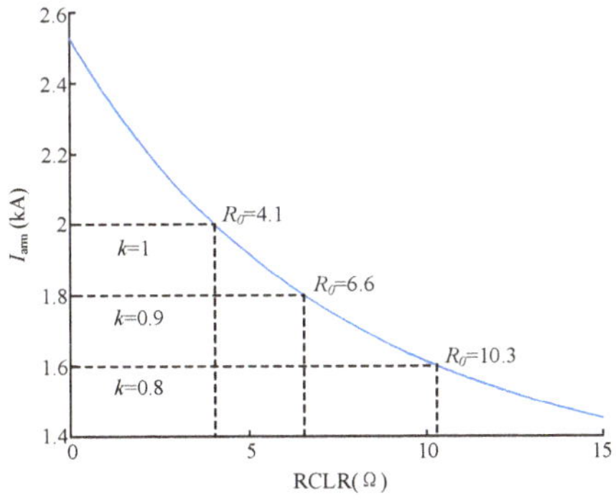

Figure 9. The maximum calculated arm currents with the different RCLR.

Take $k = 1$, $k = 0.9$ and $k = 0.8$ for examples to verify the effectiveness of the proposed calculation method. Figure 10 shows the DC currents on the faulty line where 'k = ×' means without RCLR (only using the CLI). Figure 11 shows the arm currents in MMC1 under different current limiting levels.

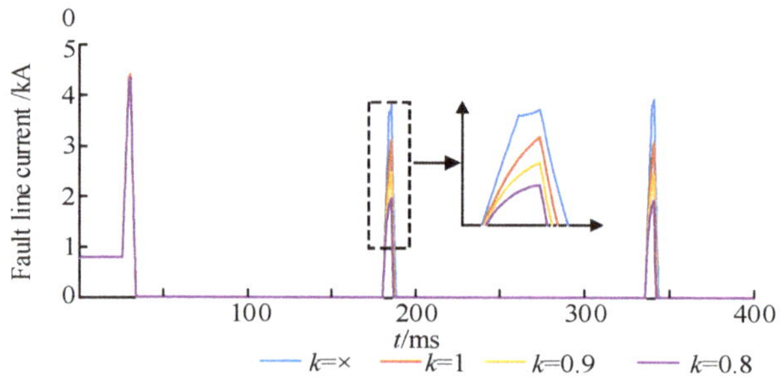

Figure 10. MMC1 DC fault current in $line_{12}$.

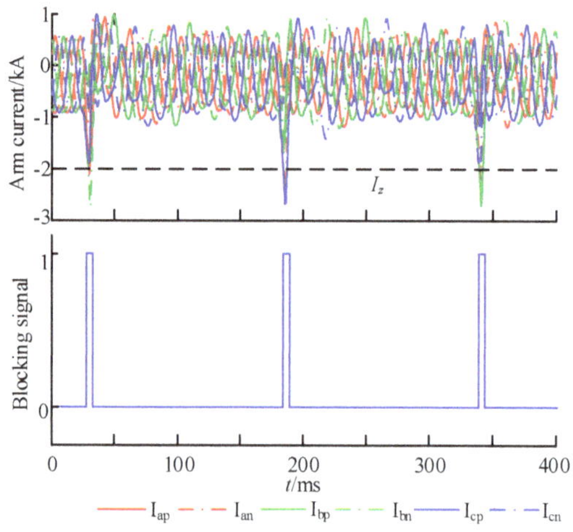

Figure 11. Arm current and blocking signal without RCLR.

As is observed in Figures 10 and 11, serious overcurrent still exists without RCLR during reclosing, and the converter is blocked again when reclosing the DCCB leading to the interruption of power flow for a long time. However, by using the proposed RCLR, the fault currents in both the faulty line and converter will be reduced significantly during reclosing, and the converter can keep continuous operation without blocking. As it is observed in Figure 12, the maximum value of arm current under different current limiting level is close to but less than the specified value and the converters are not blocked during the reclosing process. The proposed method can meet the overcurrent limiting requirement with a lower resistance value proving the calculation method is accurate and effective.

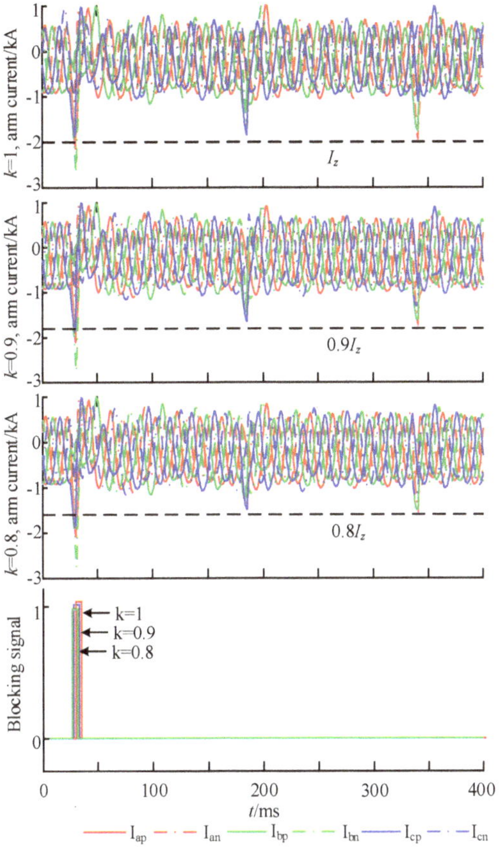

Figure 12. Arm current and blocking signal with RCLR.

5. Discussion

There has been wide research on the fault current limiter (FCL), such as the current limiting inductor (CLI), superconducting fault current limiter (SFCL), Solid State FCL and resistive fault current limiter. All the limiters will work when the fault occurs to limit the current to a lower amplitude and then break the fault circuit. For CLI, it is capable of limiting the overcurrent rise, but too many inductors in the DC grid will influence the transient response, stability and the speed for DCCB interrupting the fault current. Moreover, a large value inductor will bring large size and manufacturing burdens. For SFCL, it can reduce the fault current peak and the size of CLI, but it costs too much. A solid state FCL can interrupt the DC fault current very quickly in even hundreds of microseconds but the extra large amount of power electronic components bring much conduction loss. Resistive fault current limiter usually aims to accelerate the process of fault current decaying and is often activated by electronic switches but neglects the high voltage exerted on the electronic components at the switching. The above current limiting strategies all mean to limit the current to some level making the DCCB or other breakers to interrupt the fault current more quickly and easily. None of them pay attention to the post-fault process.

Research on the current limiting during the reclosing process for MMC-HVDC is really rare. Reclosing is necessary for an MMC-HVDC system to determine whether the fault is temporary or permanent. For a permanent pole-to-pole fault, the arm will undergo overcurrent again during each reclosing process. So, the overcurrent suppression during the reclosing process is of great significance. In the present literature, resistive limiters have been proposed but there are two problems remaining unsolved. Firstly, none of them accurately calculated the overcurrent, which is the base of current limiting. Secondly, a detailed configuration method of the limiter, especially of the calculation of the resistor, was not given or well-thought-out. In this paper, accurate current is calculated based on the proposed mesh current method, and the detailed calculation method of RCLR is conducted. The selection of resistor is based on different limiting levels of the fault current which is flexible for different thresholds of SMs overcurrent protection. The proposed method in this paper can limit the overcurrent to a required level with a lower resistance value.

6. Conclusions

Overcurrent will recur during the reclosing process without any current limiting strategy under a permanent pole-to-pole fault in MMC-HVDC which may threaten the safety of the power electronic elements and even the system. Moreover, the overcurrent will make the converter block repeatedly. To configure a reasonable FCL, the key is to calculate the overcurrent accurately and figure out the tolerance for the overcurrent. In this paper, a mesh current method (MCM) is proposed to calculated the short-circuit current based on the state equation. Based on the current calculation and the threshold of SMs overcurrent protection, the configuration principle of the reclosing current limiting resistance is proposed. The simulations constructed in PSCAD/EMTDC prove that the proposed method in this paper can limit the overcurrent in the arm to a required level under different limiting levels to protect the power electronic elements and preventing the converter frequently blocking during the reclosing process which may shorten the recovery time for the system. So, the overcurrent suppression strategy during the reclosing process of MMC-HVDC proposed in this paper is effective from the perspective of security and operation efficiency.

Author Contributions: B.J. defined the problem, proposed the FCL, conducted the simulation and contributed in the paper writing. Y.G. gathered the necessary data and handled the paper revisions.

Funding: This research received no external funding.

Conflicts of Interest: The authors declare no conflict of interest.

References

1. Jiang, B.; Wang, Z. The key technologies of VSC-MTDC and its application in China. *Renew. Sustain. Energy Rev.* **2016**, *62*, 297–304.
2. Jiang, B.; Gong, Y.; Li, Y. Fault Detection and Location of IGBT Short-Circuit Failure in Modular Multilevel Converters. *Energies* **2018**, *11*, 1492. [CrossRef]
3. Zhang, Y.; Ravishankar, J.; Fletcher, J.; Li, R.; Han, M. Review of modular multilevel converter based multi-terminal hvdc systems for offshore wind power transmission. *Renew. Sustain. Energy Rev.* **2016**, *61*, 572–586. [CrossRef]
4. Han, X.; Sima, W.; Yang, M.; Li, L.; Yuan, T.; Si, Y. Transient Characteristics Under Ground and Short-Circuit Faults in a ±500 kV MMC-Based HVDC System with Hybrid DC Circuit Breakers. *IEEE Trans. Power Del.* **2018**, *33*, 1378–1387. [CrossRef]
5. Zhang, L.; Zou, Y.; Yu, J.; Qin, J.; Vittal, V.; Karady, G.G.; Shi, D.; Wang, Z. Modeling, control, and protection of modular multilevel converter-based multi-terminal HVDC systems: A review. *CSEE J. Power Energy Syst.* **2017**, *3*, 340–352. [CrossRef]
6. Cwikowski, O.; Wickramasinghe, H.R.; Konstantinou, G.; Pou, J.; Barnes, M.; Shuttleworth, R. Modular multilevel converter dc fault protection. *IEEE Trans. Power Del.* **2018**, *33*, 291–300. [CrossRef]
7. Stumpe, M.; Ruffing, P.; Wagner, P.; Schnettler, A.; Schnettler, A. Adaptive single-poleauto-reclosing concept with advanced dc fault current control for full-bridge MMC VSC systems. *IEEE Trans. Power Del.* **2018**, *33*, 321–329. [CrossRef]
8. Sneath, J.; Rajapakse, A.D. Fault Detection and Interruption in an Earthed HVDC Grid Using ROCOV and Hybrid DC Breakers. *IEEE Trans. Power Del.* **2016**, *31*, 973–981. [CrossRef]

9. Mokhberdoran, A.; Carvalho, A.; Silva, N.; Leite, H.; Carrapatoso, A. Application study of superconducting fault current limiters in meshed HVDC grids protected by fast protection relays. *Electr. Power Syst. Res.* **2017**, *143*, 292–302. [CrossRef]
10. Lv, C.; Tai, N.; Jin, Z.; Zheng, X. Research on application of superconducting fault current limiter in MMC-MTDC. *J. Eng.* **2017**, *13*, 1307–1311. [CrossRef]
11. Larruskain, D.M.; Iturregi, A.; Abarrategui, O.; Etxegarai, A.; Zamora, I. Fault Current limiting in VSC-HVDC systems with MMC converters. In Proceedings of the 2016 International Symposium on Power Electronics, Electrical Drives, Automation and Motion (SPEEDAM), Anacapri, Italy, 22–24 June 2016; pp. 332–337.
12. Xie, S.J.; Qiu, Y.F.; Bi, T.S. Resistive DC fault current limiter. *J. Eng.* **2017**, *13*, 1682–1685.
13. Li, J.; Ding, X.B.; Liu, S.; Chen, Z.H.; Li, Z.H. DC fault current-limiting and isolating technique for the MMC-based DC system. In Proceedings of the 2017 China International Electrical and Energy Conference (CIEEC), Beijing, China, 25–27 October 2017; pp. 543–548.
14. Wang, Y.; Yuan, Z.; Wen, W.; Ji, Y.; Fu, J.; Li, Y.; Zhao, Y. Generalised protection strategy for HB-MMC-MTDC systems with RL-CFL under dc faults. *IET Gener. Transm. Distrib.* **2018**, *12*, 1231–1239. [CrossRef]
15. Li, X.; Liu, W.; Song, Q.; Rao, H.; Xu, S. An enhanced MMC topology with DC fault ride-through capability. In Proceedings of the 39th Annual Conference of the IEEE Industrial Electronics Society (IECON 2013), Vienna, Austria, 10–13 November 2013; pp. 6182–6188.
16. Sanusi, W.; Hosani, M.A.; Moursi, M.E. A novel dc fault ride-through scheme for MTDC networks connecting large-scale wind parks. *IEEE Trans. Sustain. Energy* **2017**, *8*, 1086–1095. [CrossRef]
17. Ji, S.; Wang, S.L.; Liu, T.Q. A Soft Reclosing Model for Hybrid DC Circuit Breaker in VSC-MTDC System. In Proceedings of the 2018 IEEE 4th Southern Power Electronics Conference (SPEC), Singapore, 10–13 December 2018; pp. 1–5.
18. Li, Y.; Gong, Y.; Jiang, B. A novel traveling-wave-based directional protection scheme for MTDC grid with inductive DC terminal. *Electr. Power Syst. Res.* **2018**, *157*, 83–92. [CrossRef]
19. Cwikowski, O.; Wood, A.; Miller, A.; Barnes, M.; Shuttleworth, R. Operating DC Circuit Breakers with MMC. *IEEE Trans. Power Del.* **2018**, *33*, 260–270. [CrossRef]
20. Kontos, E.; Rodrigues, S.; Pinto, R.T.; Bauer, P. Optimization of limiting reactors design for DC fault protection of multi-terminal HVDC networks. In Proceedings of the Energy Conversion Congress and Exposition (ECCE), Pittsburgh, PA, USA, 14–18 September 2014; pp. 5347–5354.
21. Kontos, E.; Bauer, P. Reactor design for DC fault ride-through in MMC-based multi-terminal HVDC grids. In Proceedings of the Annual Southern Power Electronics Conference (SPEC), Auckland, New Zealand, 5–8 December 2016; pp. 1–6.
22. Tzelepis, D.; Dysko, A.; Fusiek, G.; Nelson, J.; Niewczas, P.; Vozikis, D.; Orr, P.; Gordon, N.; Booth, C.D. Single-ended differential protection in MTDC networks using optical sensors. *IEEE Trans. Power Del.* **2017**, *32*, 1605–1615. [CrossRef]

applied sciences

MDPI

Article

Assessment of Appropriate MMC Topology Considering DC Fault Handling Performance of Fault Protection Devices

Ho-Yun Lee, Mansoor Asif, Kyu-Hoon Park and Bang-Wook Lee *

Department of Electronic engineering, Hanyang University, Hanyangdaehak-ro 55, Ansan 15588, Korea; hoyun05@hanyang.ac.kr (H.-Y.L.); mansoor1991@hanyang.ac.kr (M.A.); herochin@hanyang.ac.kr (K.-H.P.)
* Correspondence: bangwook@hanyang.ac.kr; Tel.: +82-031-400-4752

Received: 13 September 2018; Accepted: 2 October 2018; Published: 6 October 2018

Abstract: The eventual goal of high-voltage direct-voltage (HVDC) systems is to implement HVDC grids. The modular multilevel converter (MMC) has been identified as the best candidate for the realization of an HVDC grid by eliminating the shortcomings of conventional voltage source converter (VSC) technology. The related research has focused on efficient control schemes, new MMC topologies, and operational characteristics of an MMC in a DC grid, but there is little understanding about the fault handling capability of two mainstream MMC topologies, i.e., half bridge (HB) and full bridge (FB) MMCs in combination with an adequate protection device. Contrary to the existing research where the fault location is usually fixed (center of the line), this paper considered a variable fault location on the DC line, so as to compare the fault interruption time and maximum fault current magnitude. From the point of view of fault interruption, AC and DC side transient analyses were performed for both MMC topologies to suggest the appropriate topology. The simulation result confirmed that the fault handling performance of an HB-MMC with a DC circuit breaker is superior due to the smaller fault current magnitude, faster interruption time, lower overvoltage magnitude, and lesser stresses on the insulation of the DC grid.

Keywords: half bridge (HB); full bridge (FB); modular multilevel converter (MMC); hybrid HVDC breaker (HCB)

1. Introduction

Implementing DC grids having enhanced controllability and lower energy losses could be a solution for integrating renewable energy sources that have an inherently intermittent nature. In particular, voltage source converter (VSC) technology has been identified as the best candidate for realizing DC grids due to its ability to control the AC voltage magnitude, phase angle, and output frequency. With the application of pulse width modulation (PWM) control, VSCs can offer higher response time and lower harmonic content [1–3]. However, due to the absence of current zero and high di/dt resulting from DC faults, fault handling is a serious issue faced by DC transmission lines and its solution is urgently required to further the implementation of DC grids [4,5].

Recently, modular multilevel converters (MMCs) have eliminated the shortcomings of conventional VSC topologies. With the use of a modular structure, it has been possible to achieve very high levels of voltage, which are desirable for bulk power transmission [6,7]. As for MMCs, the efficient control schemes and operational characteristics of MMCs in a DC grid have been a subject of interest in academia and industry, and many solutions have been presented [8–10]. However, DC fault and its effective handling have been identified as one of the most serious challenges in the implementation of VSC-type DC transmission systems [11,12]. The impact of DC fault location on fault interruption time and maximum fault current magnitude has not been considered previously [13,14]. The search for the best MMC topology, control schemes, protection devices, and the combination of all or some of these factors, considering the nature of the DC fault, is the subject of this paper.

A half bridge (HB) MMC necessarily requires a DC breaker because of its lack of DC fault blocking capability. Therefore, a hybrid circuit breaker (HCB), which can cope with the expected rapid rise of the fault current in an HVDC grid, was used in conjunction with an HB-MMC transmission line [15,16]. This combination is referred to as Case-1 in the rest of the paper.

On the contrary, the reverse voltage phenomenon in a full bridge (FB) MMC can limit the fault effectively within a few milliseconds. Therefore, a DC circuit breaker (DCCB) is not required. A residual circuit breaker (RCB) or an ultra-fast disconnector is used to completely remove the residual fault current and isolate the fault. This combination is referred to as Case-2 in the following sections of the paper [17].

In this paper, the fault handling performance of two mainstream MMC topologies, including HB- and FB-MMCs in combination with the DC side circuit breaker, were investigated. Contrary to the existing research, in which the fault location is usually fixed (center of the line), this paper considered a variable fault location on the DC line, so as to compare the fault interruption time and maximum fault current magnitude. Since a power system must bear the current and the voltage stress during fault occurrence and recovery, the current and voltage transients on the AC and DC sides of the power system were thoroughly analyzed.

This paper is organized as follows. The operation characteristics of the MMC and simulation model are explained in Sections 2 and 3. The interruption performance of the HB- and FB-MMC DC grids is presented in Section 4. The transient performance of the system in DC faults is presented in Section 5. The strengths and weaknesses of the two cases and the suitable MMC grid topology from the point of view of fault handling performance are proposed. The conclusion is presented in Section 6.

2. The Operation Characteristics of an MMC System

2.1. DC Fault Analysis of an MMC Before Blocking

The equivalent circuit representation of an MMC composed of an HB or FB configuration prior to the fault detection is shown in Figure 1. In this period, the DC fault current i_z has not yet reached the fault detection limit. Considering KVL, the upper loop can be written as:

$$V_x = L_x \frac{di_{xg}}{dt} + R_i i_{xg} + L_{arm} \frac{di_{xu}}{dt} + R_{arm} i_{xu} - v_{xu} + L_{sc} \frac{di_{xz}}{dt} + R_{sc} i_z + v_{nN} \tag{1}$$

where $x = a, b, c$ and L_{sc} and R_{sc} represent the equivalent values of the inductance and resistance of the transmission line.

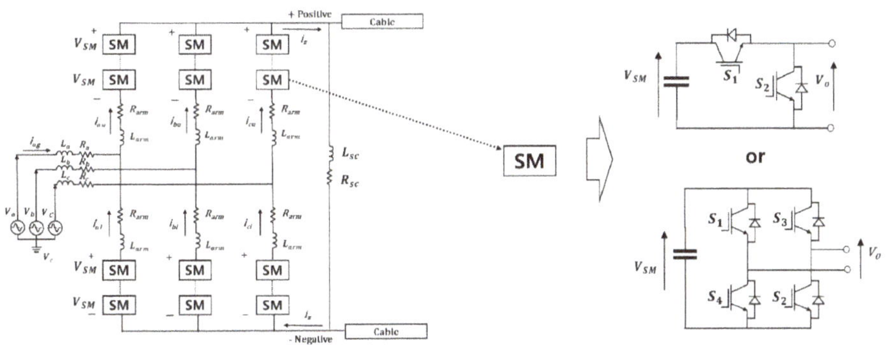

Figure 1. Equivalent circuit of a modular multilevel converter (MMC) system composed of a half bridge (HB) or full bridge (FB) configuration.

The lower loop for each phase can be represented by:

$$V_x = L_x \frac{di_{xg}}{dt} + R_i i_{xg} - L_{arm} \frac{di_{xl}}{dt} - R_{arm} i_{xl} - v_{xl} + v_{nN} \quad (2)$$

Also applying KCL, in each node we obtain:

$$i_x = i_{xu} - i_{xl} \quad (3)$$

$$i_z = i_{al} + i_{bl} + i_{cl} \quad (4)$$

$$i_z = i_{au} + i_{bu} + i_{cu} \quad (5)$$

Subtracting Equation (2) from (1) yields:

$$L_x \frac{d(i_{xu} + i_{xl})}{dt} + R_z (i_{xu} + i_{xl}) + L_{sc} \frac{di_z}{dt} + R_{sc} i_z = (v_{xu} + v_{xl}) \quad (6)$$

Equation (6) is valid for all the phases. Adding Equation (6) for each phase, we obtain:

$$(2L_z + 3L_{sc}) \frac{di_z}{dt} + (2R_z + 3R_{sc}) i_z = \sum_{x=a,b,c} (v_{xu} + v_{xl}) \quad (7)$$

For an MMC at any instant of time, N number of submodules are switched on, where N is the number of submodule (SM) per arm. Therefore, it can be written that:

$$(V_{xu} + V_{xl}) = N V_{sm} \quad (8)$$

According to Reference [18], at this point in time all the capacitors are parallel and the DC current can be expressed as:

$$i_z = \frac{2C_{sm}}{N} \sum_{x=a,b,c} (v_{xu} + v_{xl}) \quad (9)$$

where C_{sm} is the capacitance of each individual SM. Substituting Equation (9) into Equation (7) will give:

$$(2L_z + 3L_{sc}) \frac{d^2 i_z}{dt} + (2R_z + 3R_{sc}) \frac{di_z}{dt} + \frac{N}{2C_{sm}} i_z = 0 \quad (10)$$

Equation (10) provides the equation of the DC current i_z. This equation will be valid until i_z reaches the limit [19].

2.2. DC Fault Analysis of an MMC after Blocking

In an HB-MMC, as soon as a fault is detected, the IGBTs are blocked for their protection, but the anti-parallel freewheeling diodes still provide a path for the AC current to the DC line, thus feeding the fault. Therefore, DCCB is essentially required to block the fault current.

An FB-MMC system has the capability of suppressing the fault current. Initially, a current surge is allowed to flow through the IGBTs and feed the DC fault. However, as soon as the IGBTs are blocked, there is only one available current path through the series and reverse-connected DC capacitors of the submodules.

After the IGBTs are blocked, the fault current passes through the freewheeling diodes. In the process, the capacitors of FB submodules develop opposite polarity compared to the fault current during the DC fault. The complete blocking of the DC fault current takes place when the total voltage of each submodule capacitor becomes higher than the maximum peak line-to-line AC voltage, as shown in Equation (11). As a result, the fault current can be limited and a DC breaker with a low current rating can be used to isolate the faulty transmission line.

$$V_{AC,max} < V_{arm,a} + V_{arm,b}, \tag{11}$$

where $V_{arm,a}$ and $V_{arm,b}$ represent the arm voltages which are stacked at each arm through the fault path.

The operation of FB- and HB-MMCs under DC side faults have been extensively covered in References [10,19,20].

3. Simulation Model

3.1. Test Bed Model

To compare the fault handling performance between HB- and FB-MMCs, a test bed was modeled in Matlab/Simulink, as shown in Figure 2. A bipolar point-to-point HVDC link was modeled using 40-level HB- and FB-MMCs. In the steady state, the MMC was operated with constant active and reactive power control. The reference voltage of the DC link was set at 80 kV, and the length of the transmission line was 100 km. The pole-to-pole fault was generated at 0.2 s.

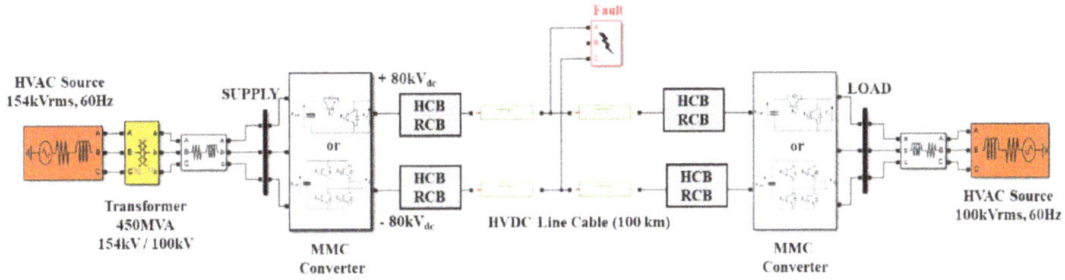

Figure 2. Test bed model in Matlab/Simulink. Converters are composed of 40-level HB- and FB-MMCs. The pole-to-pole DC fault is introduced at a variable distance from the rectifier.

Table 1 presents a summary of the HVDC system parameters. After the occurrence of the DC fault, the main controller trips the HB and FB converters within 500 microseconds to protect the converter. By performing this process, the DC fault contribution from the submodule capacitor can be prevented in HB-MMCs. The configuration of submodules in FB-MMCs does not allow the discharge of the DC side capacitor.

Table 1. Summary of high-voltage direct-voltage (HVDC) system parameters.

Parameters	Specifications
Voltage source converter (VSC) HVDC type	Bipolar HB-/FB-MMC
AC source voltage (rectifier side)	154 kV
Number of submodules per arm	40
Equivalent capacitance	10 uF
Current-limiting reactor	20 mH
Transformer power rating	450 MVA
Transformer voltage ratio	154 kV/100 kV
DC cable resistance	0.0133 Ω/km
DC cable inductance	0.8273 mH/km
DC cable capacitance	0.0139 uF/km
Length of transmission line	100 km

3.2. Comparison of System Operating Characteristics and Necessary Components by Case Types

The converter operation of HB- and FB-MMCs, after the fault detection, is shown in Figure 3. In the case of the HB-MMC, even if the IGBT is blocked, the converter continues to feed fault due to the presence of the freewheeling diode, so a DCCB is required to break the fault current. Due to the large fault current magnitude and its steep slope, a reactor is added in series with the DC line to limit the maximum stress on DCCB and assist its fault interruption operation. In the absence of a large reactor, DCCB is unable to perform fault interruption.

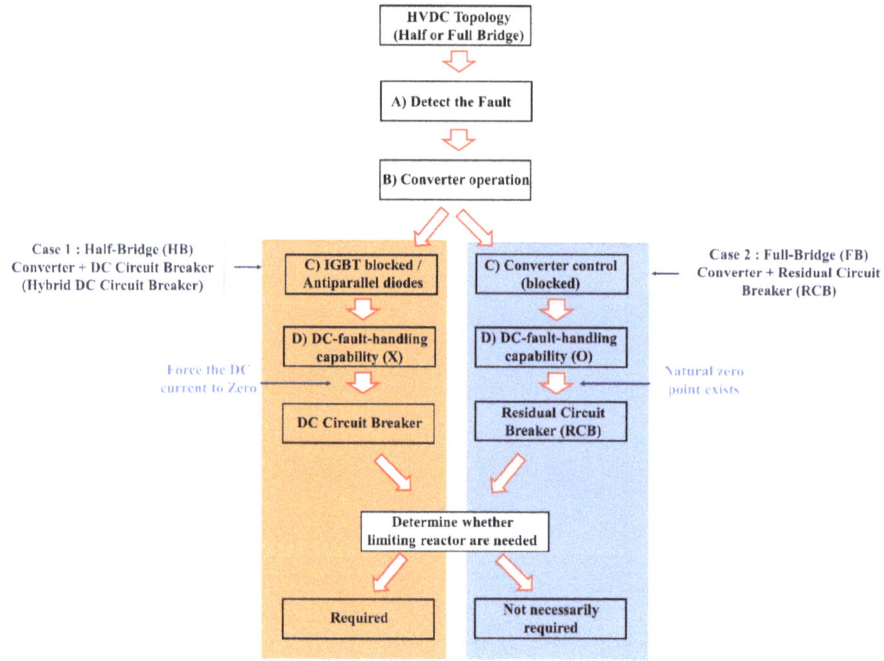

Figure 3. Comparison of overall operating characteristics and the need for a current-limiting reactor.

In the case of the FB-MMC, all IGBTs are blocked in the same way as the HB-MMC, but a reverse voltage is generated in the fault current path, so it is possible to cut off the current of the AC source by the converter itself. Therefore, when the natural zero point is generated through the converter blocking process, the DC fault is completely isolated through the RCB. Therefore, the FB-MMC system does not require a DCCB. As a result, a large limiting reactor is not necessarily required. The benefit of a limiting reactor, considering to its cost and size, is minimal. However, if the reactor is considered, the peak value of fault current will be reduced and time to reach the peak will increase. Due to its limited advantage, the limiting reactor can be left off in FB-MMC-based DC grids [21].

The operation characteristics of the two types of systems and the need for a current-limiting reactor are reflected in the simulation.

3.2.1. Case-1: Fault Management in an HB-MMC DC System Through DCCB

For HB-MMC systems, DCCBs are a necessary requirement for effectively clearing the DC faults, as shown in Figure 4a.

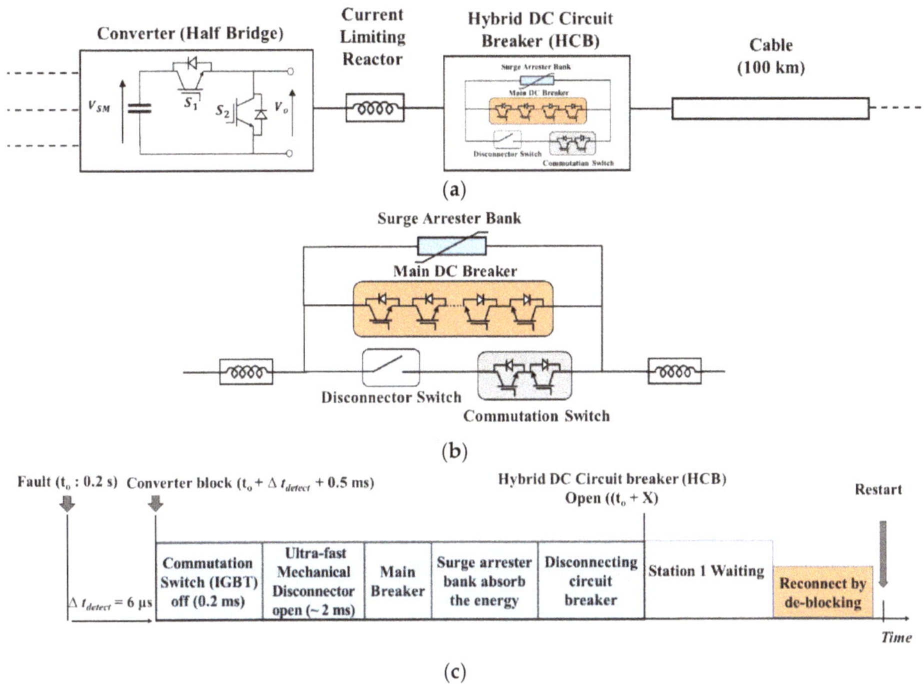

Figure 4. Case-1: DC fault handling solution for an HB-MMC based on a hybrid circuit breaker (HCB): (**a**) configuration of Case-1; (**b**) structure of the HCB; (**c**) fault handling under a DC short-circuit fault.

Unlike AC circuit breakers (ACCBs), HVDC circuit breakers (CBs) must create a current zero and dissipate the energy stored in the DC network. Recently, industry and academia have extensively researched DCCBs to overcome various well-known limitations. Among the developed prototypes, HCBs, which can block a 9-kA fault current within 5 ms, are attracting attention as the most suitable circuit breakers for application in HVDC systems [22,23].

In this paper, we proposed an HCB due to its excellent interruption capability (large di/dt with large dv/dt) and small interruption times (<5 ms). In addition, a large current-limiting reactor was installed at the DC terminal of the MMC to assist the operation of the HCB.

Figure 4b presents the structure of the HCB mentioned above. The HCB contains a load commutation switch (LCS), an ultrafast mechanical disconnector (UFMD), and a main breaker (MB). As shown in the operation characteristics of Figure 4c, the detection time of the fault current is assumed to be 6 μs. The converter is blocked within 0.5 ms after the detection of a fault. During the normal operation of the HCB, the load current flows through a nominal path that comprises of LCS and UFMD. When a DC fault is detected, the LCS opens immediately and the current is commutated to the MB. Then the UFMD opens within 2 ms and isolates the LCS from the faulted line. With the UFMD in an open position, the commutation path interrupts the fault current, and energy of the fault is absorbed by the arrestor bank. In this process, the DC current decreases quickly and can be cleared within a few milliseconds.

3.2.2. Case-2: Fault Management in an FB-MMC Using RCB

FB-MMCs are a superior alternative to HB-MMCs in terms of fault handling capability. As shown in Figure 5a, the rapid reverse of the DC voltage control of an FB-MMC can be typically provided within a few milliseconds; therefore, DCCB are not necessarily needed for protection.

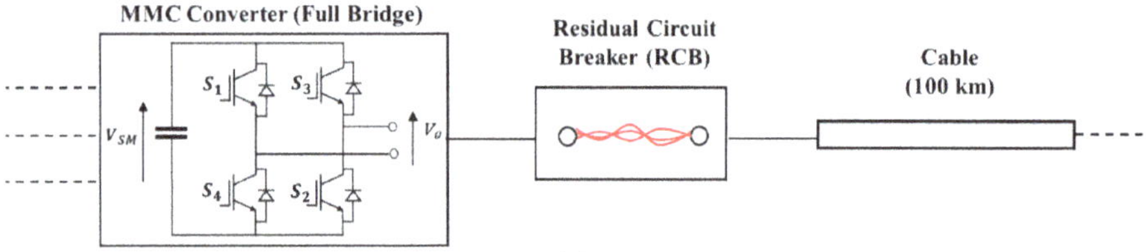

(a)

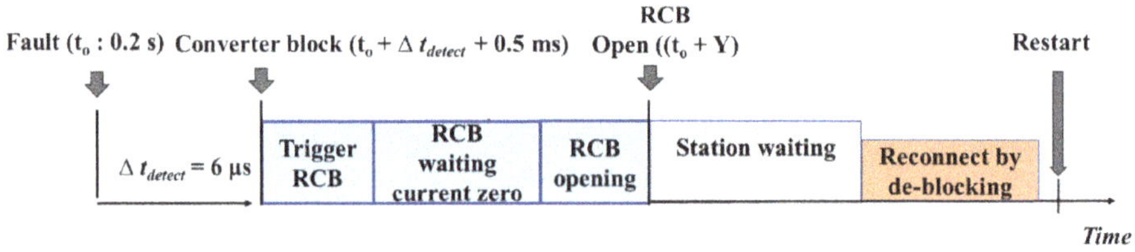

(b)

Figure 5. Case-2: DC fault handling solution composed of an FB-MMC with a residual circuit breaker (RCB): (**a**) configuration of Case-2; (**b**) operation process for DC fault handling under a DC short-circuit fault.

When the fault is detected, all IGBTs of the SMs are blocked, and the capacitors generate a reverse voltage to block the AC side currents. Since the converter has already interrupted the fault current, an RCB is sufficient to isolate the fault.

Also, unlike Case-1, the FB-MMC system does not require a DCCB. Figure 5b shows the fault handling mechanism in the FB-MMC system: detecting the fault, blocking the converter, and triggering the RCB to generate the natural current zero. Finally, if a reconnection is established by deblocking the converter and reclosing the RCB, system can be restored to the normal state.

4. Interruption Performance of HB- vs FB-MMC DC Grids

4.1. Comparison of Fault Current Blocking Performance (Pole-to-Pole Fault)

Prior to the analysis of fault handling performance with protection devices, it is important to analyze the fault current blocking ability of the two types of MMCs. We simulated a pole-to-pole fault at 0.2 s on the DC side of the rectifier to consider the worst-case fault.

Figure 6a shows that the peak current magnitude is about 30 kA for both topologies. In the case of the HB-MMC system, a fault current of about 5 kA is sustained due to the presence of freewheeling diodes.

However, in the case of the FB-MMC system, all IGBTs are blocked when a fault is detected, and a reverse voltage is generated, which forces the fault current to zero.

The DC side voltage during fault transient is shown in Figure 6b. In the case of the HB-MMC system, the voltage in the transient state suddenly reverses the polarity, and then approaches zero in about 20 ms. However, in the case of the FB-MMC system, the voltage oscillations are sustained on the DC side for about 15 ms.

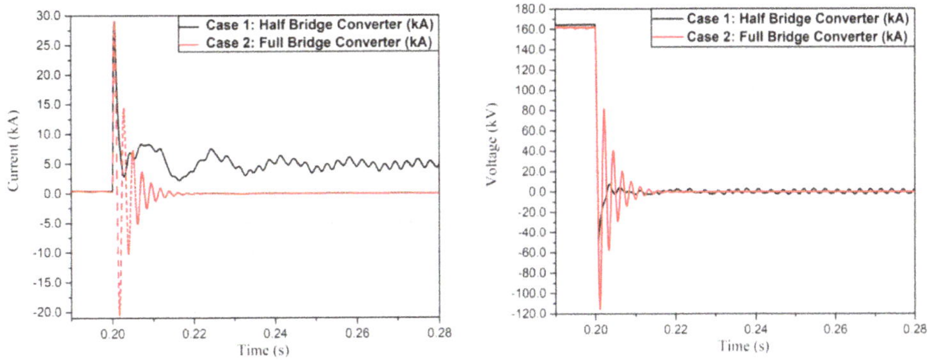

(a) (b)

Figure 6. Simulation of a DC pole-to-pole fault with MMCs: (**a**) current waveform in the transient state; (**b**) voltage waveform in the transient state.

4.2. Comparison of Fault Current Blocking Performance (Pole-to-Ground Fault)

In this section, we simulated a pole-to-ground fault at 0.2 s on the DC side of the rectifier to consider the worst-case fault. Figure 7a shows that the peak current magnitude is about 23.5 kA for both topologies. In the case of the HB-MMC system, the fault current of about 11.5 kA is sustained due to the presence of freewheeling diodes. However, in the case of the FB-MMC system, all IGBTs are blocked when a fault is detected, a reverse voltage is generated to block the fault current, and the fault current naturally approaches zero. The DC side voltage during the fault transient state is shown in Figure 7b. In both systems, voltage oscillations are sustained on the DC side.

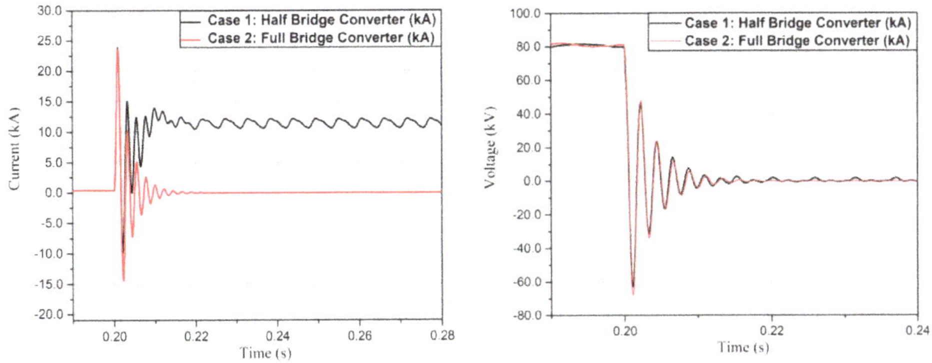

(a) (b)

Figure 7. Simulation of a DC pole-to-ground fault with MMCs: (**a**) current waveform in the transient state; (**b**) voltage measured at converter terminals.

4.3. Comparison of Fault Interruption Performance

In this section, we compared the interruption performance of Case-1 and Case-2 according to the location of the occurrence of fault on the DC line. The peak magnitude of the fault current and interruption time are considered key parameters for evaluating the performance of the two cases.

For a pole-to-pole fault at 5 km from the rectifier, the peak magnitude of the fault current in Case-2 is 15.5 kA; it is only 3.5 kA in Case-1, as shown in Figure 8. The lower peak fault current magnitude in Case-1 is due to the presence of the current-limiting reactor and the fast-operating HCB.

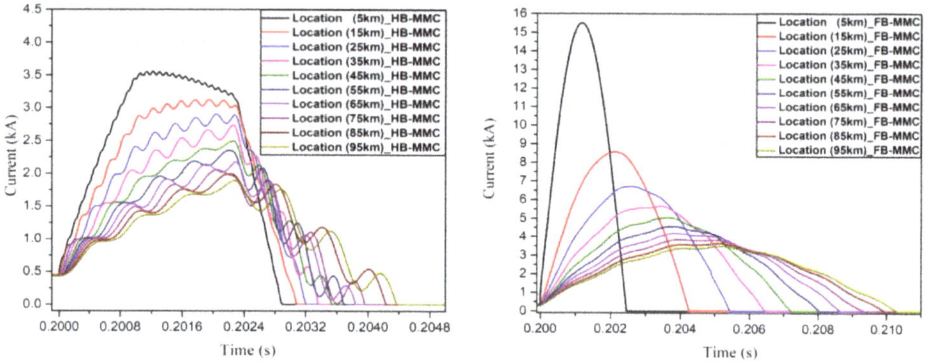

(a) (b)

Figure 8. Fault handling interruption performances: (**a**) fault current interruption characteristic of Case-1 with an HCB; (**b**) fault current interruption characteristic of Case-2 with an RCB.

The total interruption time as well as the peak fault current magnitude in the two cases is compared in Figure 9.

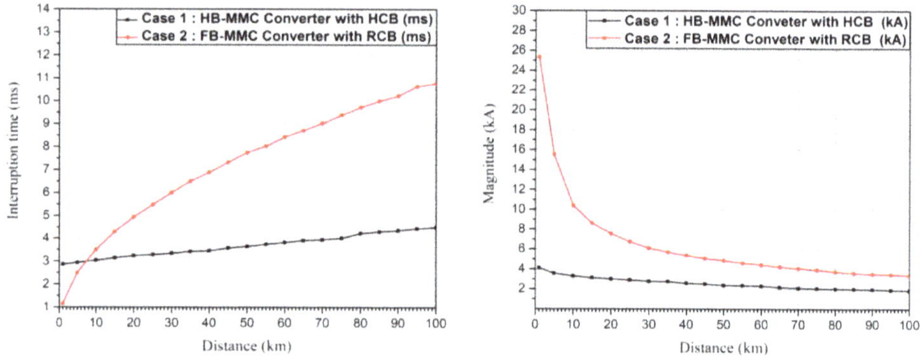

(a) (b)

Figure 9. Comparison analysis of interruption characteristics according to the location of the fault on the DC line: (**a**) interruption time; (**b**) maximum fault current magnitude.

It can be seen in Figure 9a that the interruption time in Case-1 does not change considerably with the variation in fault location. However, in Case-2 the interruption time increases considerably for faults farther away from the rectifier. The quick interruption in Case-1 can be attributed to the presence of the fast-acting HCB. However, the inherent blocking characteristics of the FB-MMC in Case-2 are dependent on the natural elimination of the fault current. Therefore, if the fault location moves away from the rectifier, the interruption time increases due to the increase in inductive and capacitive components of the transmission line. It can be seen in Figure 9a that for faults occurring beyond 8 km, Case-1 offers a smaller interruption time.

Figure 9b shows that in both cases the peak magnitude of the fault current decreases upon increasing the distance of the fault from the rectifier. However, due to presence of an HCB in Case-1, the fault is interrupted before its natural peak and therefore the peak magnitude is quite low regardless of the fault location. However, in Case-2 the peak magnitude of the fault current is dependent on the inductive reactance of the line. Therefore, the peak magnitude is quite high for faults occurring on the converter terminal.

In conclusion, we found that Case-1 has a lower fault current magnitude and lower interruption time for most fault locations.

Figure 10 shows the comparison of the energy dissipation of the circuit breaker according to the fault location.

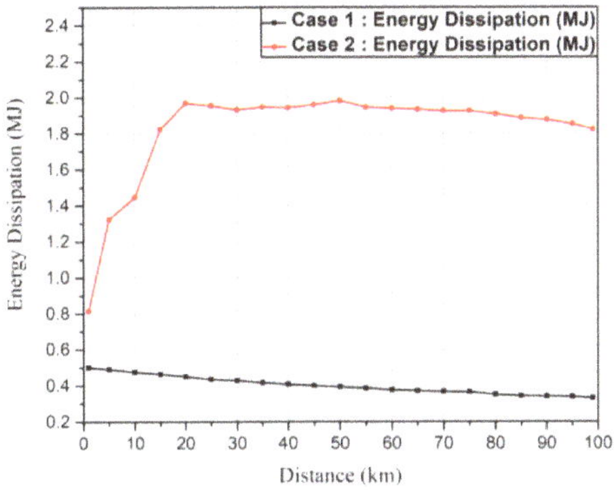

Figure 10. Comparison of energy dissipation on the circuit breaker (CB). Case-1 is an HCB; Case-2 is an RCB.

In Case-1, the circuit breaker must fully cope with the DC fault current, which has high di/dt. Therefore, we confirmed that the energy dissipation in the HCB is more than 0.32 MJ, even if the fault location is far away from the converter side.

In Case-2, the peak fault current is higher than that in Case-1 and the interruption time is also larger. Therefore, the energy dissipation in the RCB was found to be about 2 MJ when the fault occurs at 20 km.

5. Transient Performance of the System in DC Faults

The occurrence of DC faults and their clearance introduce current and voltage transients on both the AC and DC sides of the power system. These transient events affect the power quality and can potentially deteriorate the power system components. In this section, an analysis of the transient response of the power system in two cases is presented.

5.1. Analysis of the Current and Voltage Transient of the AC and DC Sides Considering a Reclosing Operation

Since, most transmission line faults are temporary, a reclosing operation is considered for a complete analysis of fault transients. The power system transients were analyzed to determine the suitable grid topology for the two cases.

The pole-to-pole fault was simulated at 0.2 s at a distance of 5 km from the rectifier, and the converter was set to operate within 500 μs after the detection of a fault. The circuit breaker attempts to clear the fault immediately after converter blocking. A delay of 0.2 s is introduced to allow for the deionization of fault location, following which the de-blocking of the converter and reclosing of the circuit breaker are initiated.

5.1.1. Transient Current and Voltage of the AC System

Figure 11 shows the AC side current and voltage waveforms for Case-1. Figure 11a presents the AC side current graph. It shows that the peak current approaches 2 kA just after 0.2 s, i.e., before fault interruption. After 0.4 s, i.e., during reclosing, the AC current exceeds its steady-state value before reaching the steady state in 50 ms. A voltage transient with a peak of up to 90 kV can be observed during the reclosing operation, as shown in Figure 11b.

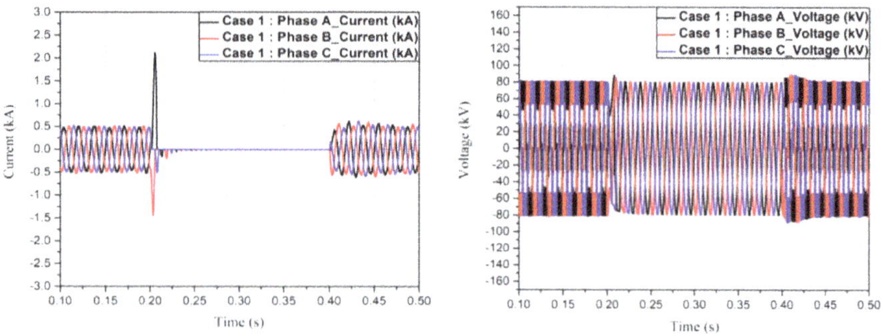

(a) (b)

Figure 11. The transient waveforms of the AC system in Case-1: (**a**) current waveform; (**b**) voltage waveform.

Figure 12 shows the AC-side current and voltage waveforms for Case-2 during interruption and reclosing operations. It can be seen in Figure 12a that there is no current overshoot during fault inception due to the instant blocking of IGBTs. During reclosing, the maximum current does not exceed −1.02 kA, which is 0.4 kA higher than that of Case-1.

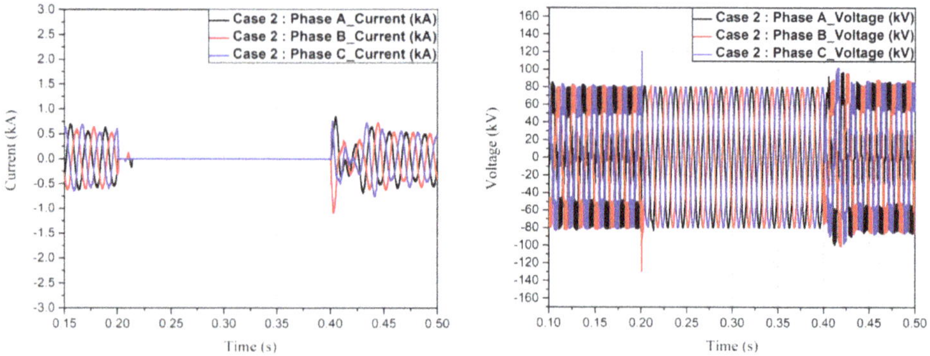

(a) (b)

Figure 12. The current and voltage waveform at the AC side of Case-2 in interruption and reclosing operations: (**a**) current waveform; (**b**) voltage waveform.

Figure 12b shows that an overvoltage of 120 kV occurs during the converter blocking at the time of fault. An overvoltage of up to 100 kV occurs during the reclosing process, followed by a return to the steady state in 10 s.

As a result, although the magnitude of the current and the voltage are not much different on the AC side, the current magnitude in the reclosing and the overvoltage magnitude in the interruption process are relatively high, and the time to achieve the steady-state voltage after the reclosing process is very long. Therefore, Case-1 would have advantages in terms of the stability of the AC system and lower stress on the insulation.

5.1.2. The Analysis of the DC Current and Voltage Waveform Across the HCB and RCB in the Transient State

Figure 13 is a graph showing the DC current flowing through the circuit breaker and the voltage across the circuit breaker during the fault and reclosing process. As shown in Figure 13a, in Case-1, the maximum fault current is limited to 3.3 kA due to the quick HCB operation.

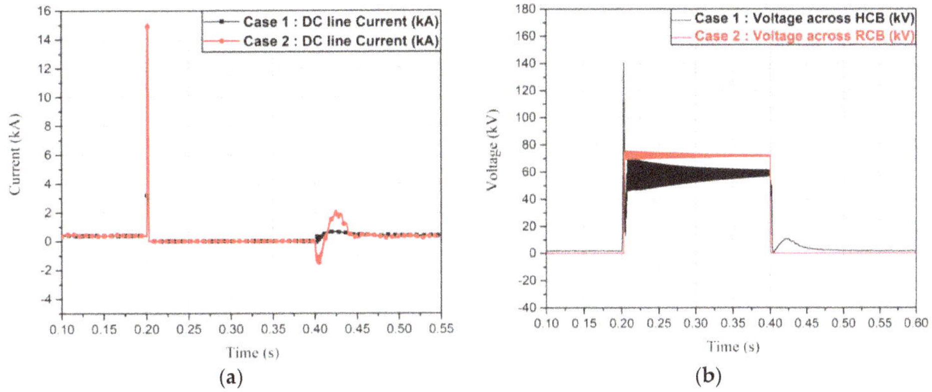

Figure 13. The waveform of the DC line current and voltage across the CB in the transient state: (**a**) the DC line current; (**b**) the voltage across the CB.

In Case-2, however, due to the fault handling capability of the FB-MMC, it is confirmed that the fault current rapidly increases to about 15.5 kA and then rapidly decreases.

In the reclosing operation, Case-1 has a maximum fault current magnitude of 0.8 kA, and it is confirmed that it achieves a steady state in 100 ms. Case-2 showed high di/dt as well as the peak magnitude, which is about 2 kA—more than twice the value of that in Case-1.

Considering the insulation design of the system and device, Case-2 with a high current magnitude and high di/dt slope would be detrimental to the power system components.

The voltage waveform across the circuit breaker is presented in Figure 13b. In Case-1 and Case 2, the voltage across the circuit breaker increased suddenly after 0.2 s. In the reclosing process, the recovery voltage across the HCB decreases within 50 ms.

5.1.3. The Transient Overvoltage Analysis During DC Faults

This section compares the overvoltage characteristics across the DC line as a result of DC faults in two cases.

The voltage across the converter in a pole-to-pole fault is presented in Figure 14. In Case-1, the voltage dip is not large due to the presence of voltage across the circuit breaker during the fault. In addition, it was confirmed that the voltage in the normal state recovered to 160 kV within 100 ms after the peak overvoltage of 175 kV occurred during the reclosing operation.

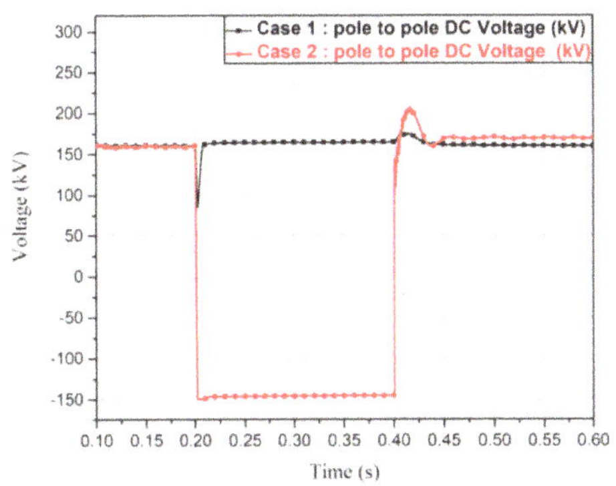

Figure 14. The waveform of pole-to-pole DC voltage in the case of pole-to-pole fault.

On the other hand, in Case-2, a reverse voltage was applied across the DC system due to the blocking operation of the FB-MMC to prevent the AC system from feeding in to the DC fault. In the case of reclosing, an overvoltage of 205 kV was obtained. In addition, the recovery time to a DC-rated voltage of 160 kV after reclosing was much longer than that of Case-1, and it was confirmed that the steady-state voltage was reached within 10 s, compared to 100 ms in Case-1.

Figure 15a shows the overvoltage graph of Case-1, showing the maximum overvoltage of 90 kV during the reclosing process. However, in Case-2, a maximum overvoltage of 150 kV and a longer time to reach the steady state was observed, as shown in Figure 15b.

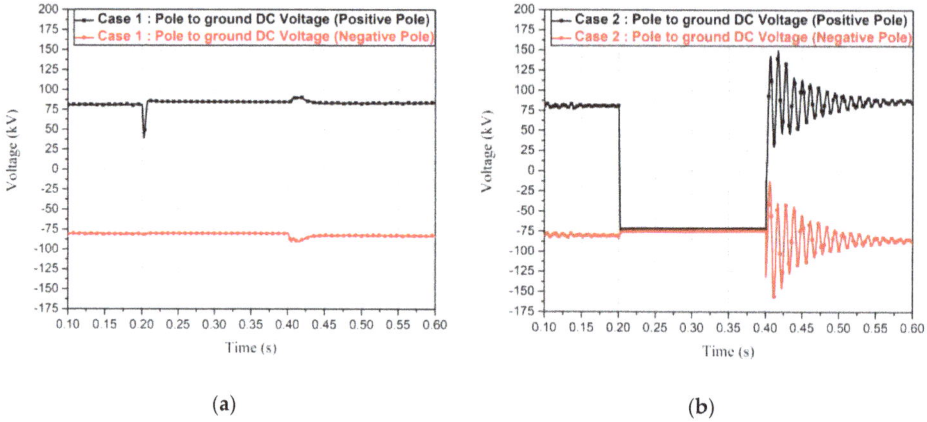

Figure 15. The waveform of pole-to-ground DC voltage in the fault and reclosing process: (**a**) Case-1; (**b**) Case-2.

System components are expected to carry larger currents as well as bear higher overvoltage stress during fault transience in Case-2.

5.2. The Preferable Case for HVDC Grid Application

In this section, we evaluated the feasibility of applying two fault handling solutions to HVDC grids based on the simulation results. The summary of our analysis is presented in Table 2.

Table 2. Comparison of the two fault-handling solutions.

Parameter	Simulation Condition	Case-1	Case-2
Structure DC-fault-handling capability Current-limiting reactor	-	HB-MMC + HCB O Required	FB-MMC + RCB O Not necessarily required
Total interruption time (ms) Maximum DC fault current in fault period (kA) Energy dissipation across circuit breaker (MJ)	Variable fault location	3~4.5 (ms) 2~4 (kA) 0.32~0.51 (MJ)	~11 (ms) 3~25 (kA) 0.01~0.05 (MJ)
Maximum AC current in fault period (kA) Maximum AC current in reclosing period (kA)	5 km from the sending end converter side	2 (kA) 0.62 (kA)	0.15 (kA) 1.02 (kA)
Maximum DC current in reclosing period (kA) Maximum DC system overvoltage (kV) Time to recovery (ms)	5 km from the sending end converter side (pole-to-pole)	0.8 (kA) 175 (kV) ~0.1 (s)	2 (kA) 205 (kV) 10 (s)
Maximum DC overvoltage in reclosing period (kV)	5 km from the sending end converter side (pole-to-ground)	91 (kV)	152 (kV)
Stress on insulation	-	Relatively low	Relatively high
Feasibility in HVDC grid application	-	★★★★★	★★☆☆☆

* Excellent ★★★★★, Very Good ★★★★☆, Good ★★★☆☆, Fair ★★☆☆☆, Poor ★☆☆☆☆

Although in Case-1 the HCB alone copes with a large DC fault current, the power system experienced lower overvoltage stress. However, considering the high energy dissipation in the HCB, a sophisticated insulation design is required.

Table 2 shows that in Case-1, the fault current, overvoltage magnitude, and interruption time are low. Conversely, in Case-2, the abovementioned parameters are relatively high. Therefore, additional consideration for the insulation design is required to ensure the reliable long-term operation of the DC grid. Furthermore, the oscillating voltage in the transient state is expected to stress the overall system components. As a result, we conclude that Case-1 would be a better solution for HVDC grid application.

6. Conclusions

Two types of representative MMC systems along with an HCB or RCB have been investigated for the feasibility of application in an HVDC grid. The comparative studies have been conducted in terms of maximum fault current, total interruption time, and voltage characteristics of the grid during the transient period.

Although the HB-MMC with a DC circuit breaker has a disadvantage of stress on the HCB associated with the insulation design of the CB itself, it appears to be the likely candidate for application in future DC grids due to its low fault current, low interruption time, low overvoltage magnitude, and faster recovery. Furthermore, it is beneficial in terms of insulation design because it applies relatively low voltage stress to various power system components during the transient period.

Author Contributions: H.-Y.L. conceptualized the topic, formulated methodology, performed simulations, and prepared an original draft; M.A. performed a formal analysis and reviewed the draft; K.-H.P. curated the data and edited the draft; B.-W.L. supervised the study.

Funding: This research received no external funding.

Acknowledgments: This work was supported by the Ministry of Trade, Industry, and Energy of Korea through the Human Resources Program in Energy Technology of the Korea Institute of Energy Technology Evaluation and Planning (KETEP) under Grant 20174030201780 and by the KEPCO Research Institute under the project entitled by "Design of analysis model and optimal voltage for MVDC distribution system (R17DA10)".

Conflicts of Interest: The authors declare no conflict of interest.

References

1. Shabestari, P.M.; Ziaeinejad, S.; Mehrizi-Sani, A. Reachability analysis for a grid-connected voltage-sourced converter (VSC). In Proceedings of the 2018 IEEE Applied Power Conference and Exposition (APEC), San Antonio, TX, USA, 4 March 2015; pp. 2349–2354.
2. Yazdani, A.; Iravani, R. *Voltage-Sourced Converters in Power Systems: Modeling, Control, and Applications*; Wiley-IEEE Press: Hoboken, NJ, USA, 2010; pp. 115–125. ISBN 978-0-470-52156-4.
3. Shewarega, F. Simplified modeling of VSC-HVDC in power system stability studies. *IFAC Proc. Vol.* **2014**, *47*, 9099–9104. [CrossRef]
4. Frank, C.M. HVDC circuit breakers: A review identifying future research needs. *IEEE Trans. Power Deliv.* **2011**, *26*, 998–1007. [CrossRef]
5. Mokhberdoran, A.; Carvalho, A. A review on HVDC circuit breakers. In Proceedings of the 3rd Renewable Power Generation Conference (RPG 2014), Naples, Italy, 24–25 September 2014.
6. Rodriguez, J. Multilevel converters: An enabling technology for high-power applications. *Proc. IEEE* **2009**, *97*, 1791–1792. [CrossRef]
7. He, Z.; Hu, J. Mechanical DC circuit breaker and FBSM-based mmc in a high-voltage MTDC networks: Coordinated operation for network riding through dc fault. In Proceedings of the Renewable Power Generation (RPG 2015), Beijing, China, 17–18 October 2015; pp. 1–6.
8. Chen, X.; Zhao, C. Research on the fault characteristics of HVDC based on modular multilevel converter. In Proceedings of the 2011 IEEE Electrical Power and Energy Conference, Winnipeg, MB, Canada, 3–5 October 2011.

9. Jonsson, T.; Lundberg, S. Converter Technologies and Functional Requirements for Reliable and Economical HVDC Grid Design. In Proceedings of the 2013 CIGRE Canada Conference, Calgary, AB, Canada, 9–11 September 2013.
10. Najmi, V. Modelling, Control and Design Considerations for Modular Multilevel Converters. Master's Thesis, Electrical Engineering Department, Virginia Polytechnic Institute and State University, Blacksburg, VA, USA, May 2015.
11. Henry, S.; Denis, A.M. Feasibility study of off-shore HVDC grids. In Proceedings of the IEEE PES General Meeting, Providence, RI, USA, 25–29 July 2010.
12. Bucher, M.K.; Walter, M.M. Options for ground fault clearance in HVDC offshore networks. In Proceedings of the 2012 Energy Conversion Congress and Exposition (ECCE), Raleigh, NC, USA, 15–20 September 2012.
13. Petino, C.; Heidemann, M. Application of multilevel full bridge converters in HVDC multiterminal systems. *IET Power Electron.* **2016**, *9*, 297–304. [CrossRef]
14. Xu, Z.; Xiao, H. DC Fault Analysis of Clearance Solutions of MMC-HVDC systems. *Energies* **2018**, *11*, 941. [CrossRef]
15. Zhang, Z. Short-Circuit Current Calculation and performance requirement of HVDC Breakers for MMC-MTDC Systems. *IEEJ Trans. Electr. Electron. Eng.* **2016**, *11*, 168–177. [CrossRef]
16. Khan, U.A.; Lee, B.W. Feasibility analysis of a novel hybrid-type superconducting circuit breaker in multiterminal HVDC networks. *Phys. C Supercond. Appl.* **2015**, *518*, 154–158. [CrossRef]
17. Think Grid. Available online: http://www.think-grid.org/fault-blocking-converters-dc-networks-1 (accessed on 20 July 2018).
18. Win, J.; Saeedifard, M. Hybrid design of modular multilevel converters for HVDC systems based on various submodule circuits. *IEEE Trans. Power Deliv.* **2015**, *30*, 385–394.
19. Nami, A.; Liang, J. Analysis of Modular Multilevel Converters with DC Short Circuit Fault Blocking Capability in Bipolar HVDC Transmission Systems. In Proceedings of the 2015 17th European Conference on Power Electronics and Applications (EPE'15 ECCE-Europe), Geneva, Switzerland, 8 September 2015; pp. 1–10.
20. Fazel, S.S. Investigation and Comparison of Multi-Level Converters for Medium Voltage Applications. Ph.D. Thesis, Berlin University, Berlin, Germany, July 2015.
21. Kontos, E.; Pinto, R.T. Impact of HVDC transmission system topology on multiterminal DC network faults. *IEEE Trans. Power. Deliv.* **2013**, *30*, 844–852. [CrossRef]
22. Tahata, K. HVDC circuit breakers for HVDC grid applications. In Proceedings of the 11th IET International Conference on AC and DC Power Transmission, Birmingham, UK, 10–12 February 2015.
23. Hafner, J.; Jacobson, B. Proactive hybrid HVDC breakers—A key innovation for reliable HVDC grids. In Proceedings of the CIGRE International Conference, Bologna, Italy, 13–15 September 2011.

applied sciences

MDPI

Article

A Study on Stability Control of Grid Connected DC Distribution System Based on Second Order Generalized Integrator-Frequency Locked Loop (SOGI-FLL)

Jin-Wook Kang [1], Ki-Woong Shin [2], Hoon Lee [1], Kyung-Min Kang [1], Jintae Kim [1] and Chung-Yuen Won [1,*]

[1] Department of Electrical and Computer Engineering, Sungkyunkwan University, Suwon 16419, Korea; kjw2171@naver.com (J.-W.K.); dlgns0520@naver.com (H.L.); innovate_k@naver.com (K.-M.K); jintae.kim75@gmail.com (J.K.)

[2] R&D Center, WINIX Inc., Shiheung 15078, Korea; sinkiwoong@naver.com

* Correspondence: woncy@skku.edu; Tel.: +82-031-290-7169

Received: 21 July 2018; Accepted: 13 August 2018; Published: 16 August 2018

Abstract: This paper studies a second order generalized integrator-frequency locked loop (SOGI-FLL) control scheme applicable for 3-phase alternating current/direct current (AC/DC) pulse width modulation (PWM) converters used in DC distribution systems. The 3-phase AC/DC PWM converter is the most important power conversion system of DC distribution, since it can boost 380 V_{rms} 3-phase line-to-line AC voltage to 700 V_{dc} DC output with various DC load devices and grid voltages. The direct-quadrature (*d*-*q*) transformation, positive sequence voltage extraction, proportional integral (PI) voltage/current control, and phase locked loop (PLL) are necessary to control the 3-phase AC/DC PWM converter. Besides, a digital filter, such as low pass filter and all pass filter, are essential in the conventional synchronous reference frame-phase locked loop (SRF-PLL) method to eliminate the low order harmonics of input. However, they limit the bandwidth of the controller, which directly affects the output voltage and load of 3-phase AC/DC PWM converter when sever voltage fluctuation, such as sag, swell, etc. occurred in the grid. On the other hand, the proposed control method using SOGI-FLL is able to do phase angle detection, positive sequence voltage extraction, and harmonic filtering without additional digital filters, so that more stable and fast transient control is achieved in the DC distribution system. To verify the improvement of the characteristics in the unbalanced voltage and frequency fluctuation of the grid, a simulation and experiment are implemented with 50 kW 3-phase AC/DC PWM converter used in DC distribution.

Keywords: DC distribution; 3-phase AC/DC PWM converter; SOGI-FLL; phase detection

1. Introduction

In recent years, a distributed power generation system using a new, renewable energy source, and an eco-friendly electric vehicle has been attracted. The production and consumption of DC power have thus inevitably increased, and studies on the energy efficiency of DC power have been actively conducted.

In order to implement a DC distribution system with the conventional AC distribution infrastructure, a high efficiency power conversion system using a power semiconductor device is required. In general, diode rectifiers have been used to convert AC power to DC power because of their relatively simple structure and operation. However, the diode rectifier is not suitable for a DC distribution system because the constant output voltage cannot be controlled when the power factor of AC input varies according to the type and size of the load. In addition, regenerative operation is not possible due to the unidirectional characteristic. Therefore, the 3-phase AC/DC pulse width modulation (PWM) converter is widely used to compensate for these drawbacks [1,2].

The 3-phase AC/DC PWM converter used for the DC distribution, as shown in Figure 1, is composed of six insulated gate bipolar transistors (IGBTs), dc-link capacitors, and an LCL filter to the reduce harmonics on the grid side. For bidirectional power control and power factor control of this system, phase angle information of the input power source is required. In many studies [3–5], a positive sequence voltage detector using an all pass filter (APF) and synchronous reference frame phase locked loop (SRF-PLL) have been widely used; however, not only do these devices have a complicated configuration, but they are also weak in lower order harmonics [6]. In addition, problems of the grid side, such as sag, swell, and frequency fluctuation directly affect the output of the 3-phase AC/DC PWM converter. As a result, it is difficult to maintain a stable condition in the DC distribution system in which various loads are repeatedly connected [7]. In order to solve the frequency variation of the grid for the DC distribution system, improved SRF-PLL method has been researched, but the unbalanced grid voltage and voltage drop condition are not taken into account [8].

Meanwhile, the second order generalized integrator (SOGI) algorithm was introduced by Prof. Rodriguez [9]. One of SOGI applications, SOGI-frequency locked loop (SOGI-FLL) has functions, such as the positive sequence voltage extraction and phase angle detection, as well as robust filtering characteristic. Therefore, phase synchronization can be performed using only the SOGI-FLL and it does not need any additional digital filters, such as the low pass filter (LPF) that is used in the conventional method [10,11].

Therefore, in this paper, we propose advanced control method using the SOGI-FLL that can be applied to a 3-phase AC/DC PWM converter for the DC distribution. The proposed control method is applied to improve the stability and transient characteristic of the DC distribution system, and it is verified that the DC distribution system achieves stable and fast transient characteristics, even though unbalanced grid voltage, frequency variation, and voltage drop occur.

This paper is organized as follows. In Section 2, we analyze the 3-phase AC/DC PWM converter, which constitutes the DC distribution system and describes the conventional method. In Section 3, we describe the phase detection and positive sequence voltage extraction while using the proposed SOGI-FLL technique. Sections 4 and 5 give the simulation and experimental results for the proposed control method based on the SOGI-FLL control, and Section 6 provides the conclusion.

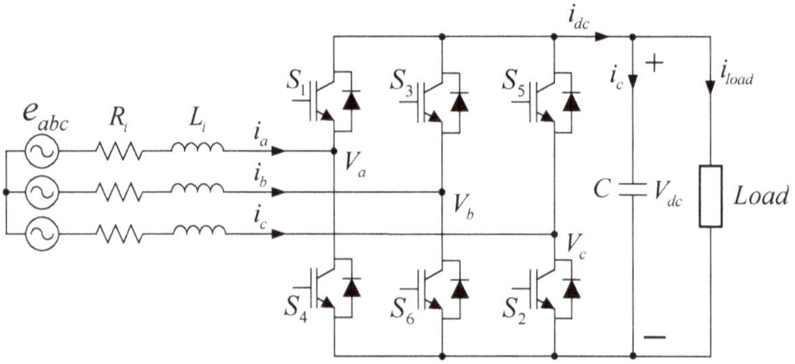

Figure 1. Configuration of 3-phase alternating current/direct current (AC/DC) pulse width modulation (PWM) converter.

2. 3-Phase AC/DC PWM Converter Used in DC Distribution System

2.1. 3-Phase AC/DC PWM Converter [12–14]

The 3-phase AC/DC PWM converter that is used for DC distribution is an AC/DC step-up converter that outputs a boost DC while using an appropriately designed AC input side reactor. Using this 3-phase AC/DC PWM converter, the grid side AC current can be maintained at a waveform that is close to a sinusoidal waveform with a relatively low total harmonic distortion (THD), and the power factor can be freely controlled. In addition, a bi-directional energy flow is possible by using a power semiconductor switch capable with bidirectional power flow, such as IGBT instead of a diode, so it can operate as both a DC/AC inverter and an AC/DC inverter.

Figure 2 shows the equivalent circuit of the 3-phase AC/DC PWM converter that is shown in Figure 1. The AC input voltage of each phase satisfies Equation (1), and the relationship between the input current i_{abc} and the voltage across the AC input reactor V_{Labc} is expressed, as shown in Equation (2).

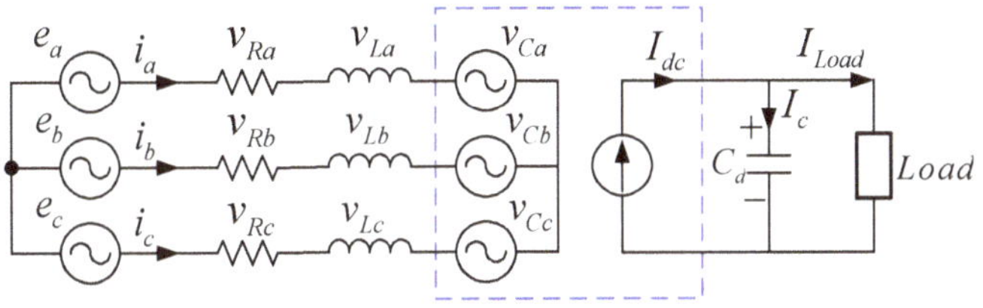

Figure 2. Equivalent circuit of 3-phase AC/DC PWM converter.

$$e_{abc} = v_{Rabc} + v_{Labc} + v_{Cabc} \tag{1}$$

$$v_{Labc} = j\omega L_{abc} i_{abc} \tag{2}$$

As can be seen from Equation (2), increasing the V_{Labc} increases the AC input side current, i_{abc}. Conversely, if the value of V_{Labc} becomes negative, the direction of i_{abc} is reversed and a regenerating operation is thus performed, the power of which flows from the DC load to the grid side. If the switching loss and harmonic loss of the 3-phase AC/DC PWM converter are ignored, the input power and the DC output power have the relationship that is shown in Equation (3).

$$V_C \cdot C \frac{dV_C}{dt} = v_{Ca} i_a + v_{Cb} i_b + v_{Cc} i_c = e_a i_a + e_b i_b + e_c i_c \tag{3}$$

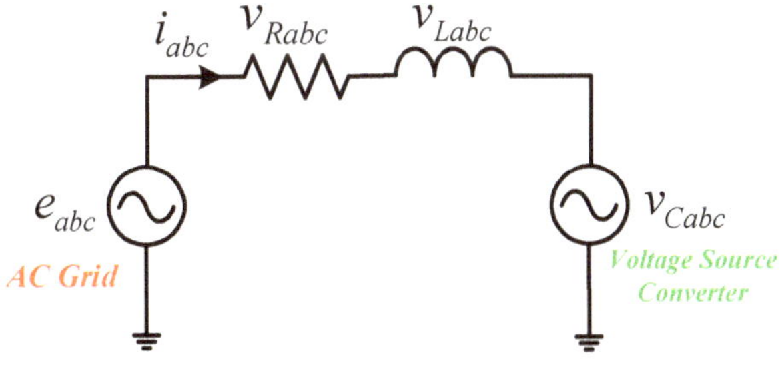

Figure 3. Equivalent circuit of a 3-phase AC/DC PWM converter.

Figure 3 shows a single-phase equivalent circuit of a 3-phase AC/DC PWM converter for analyzing the grid voltage and the current in a phasor diagram. When the converter performs a regenerative operation, the power factor angle of the AC power source in the steady state can be varied from leading to lagging, as shown in the phasor diagram of Figure 4 by controlling the AC side of the 3-phase AC/DC PWM converter. Figure 4 shows that the AC voltage is the smallest when the lagging operation is conducted. If the DC voltage is smaller than the maximum value of the AC input voltage, the constant DC output voltage can be controlled by lagging operation.

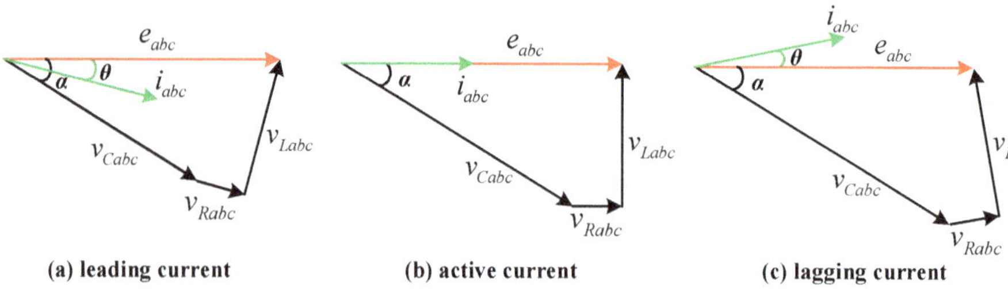

Figure 4. Phasor diagram of a 3-phase AC/DC PWM converter.

2.2. *Positive Sequence Voltage and Phase Detector [15,16]*

The phase angle information of the grid is needed in a grid-connected DC distribution system. In the case of power factor control, active/reactive power control, and harmonic current compensation, the current or voltage reference must be synchronized with the phase angle of the input voltage. A PLL is used for this phase synchronization. If the 3-phase power source is balanced, then the phase angle can be obtained by calculating the period between the zero-crossing points where the voltage of one phase or any line-to-line voltage changes from negative to positive. However, in the case of unbalanced 3-phase source, a miscalculated phase angle can result in harmonic components of voltage and current, which can lead to unforeseen problems. To avoid this problem that is caused by the unbalanced 3-phase source, the phase angle is detected by calculating the positive sequence voltage.

The equation for calculating the positive sequence voltage is given by the following Equation (4):

$$\begin{bmatrix} v_{pa} \\ v_{pb} \\ v_{pc} \end{bmatrix} = \frac{1}{3} \cdot \begin{bmatrix} 1 & \alpha & \alpha^2 \\ \alpha^2 & 1 & \alpha \\ \alpha & \alpha^2 & 1 \end{bmatrix} \cdot \begin{bmatrix} v_{as} \\ v_{bs} \\ v_{cs} \end{bmatrix} \quad (4)$$

where $\alpha = e^{j\frac{2\pi}{3}} = -\frac{1}{2} + j\frac{\sqrt{3}}{2}$ and $\alpha^2 = e^{j\frac{4\pi}{3}} = -\frac{1}{2} - j\frac{\sqrt{3}}{2}$.

α and α^2 can be substituted into Equation (4), and Equation (5) is then defined, as follows:

$$\begin{bmatrix} v_{pa} \\ v_{pb} \\ v_{pc} \end{bmatrix} = \begin{bmatrix} \frac{1}{2}v_{as} - \frac{1}{2\sqrt{3}j}(v_{bs} - v_{cs}) \\ -(v_{pa} + v_{pc}) \\ \frac{1}{2}v_{cs} - \frac{1}{2\sqrt{3}j}(v_{as} - v_{bs}) \end{bmatrix} \quad (5)$$

Figure 5 shows the control block diagram of the positive sequence voltage and PLL while using Equations (4) and (5). As shown in the figure, the *d-q* transformation is performed after extracting the positive sequence voltage while using the 3-phase voltage of the input power source. In the *d-q* transformation, the information on the *q* axis refers to the magnitude of the voltage output of the converter, and the information on the *d* axis is related to the phase error between the converter output voltage and the grid voltage. Therefore, the phase output voltage of the converter must control the *d*-axis voltage to zero for phase synchronization with the grid voltage. The *d*-axis voltage in the synchronous reference frame, as calculated by the PI controller, then performs a PLL that controls the phase angle to be zero so that the phase angle matches the actual phase angle of the grid side. When the input power source includes harmonics due to noise, the harmonics also appear in the *d*-axis voltage used as an input of the PLL. Therefore, as shown in Figure 6, the first order low pass filter is used to prevent pulsation.

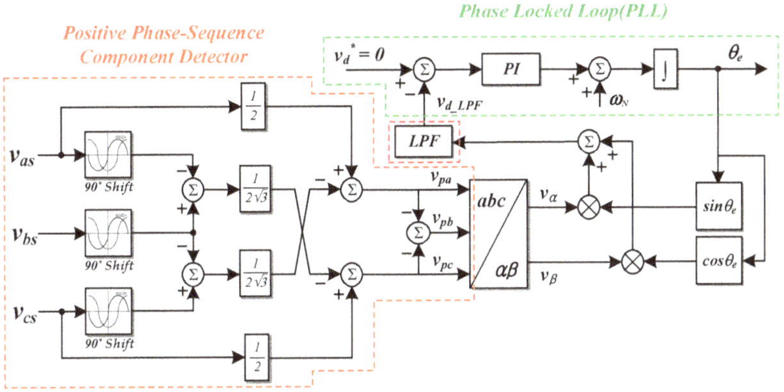

Figure 5. Control block diagram of positive sequence voltage and phase detector.

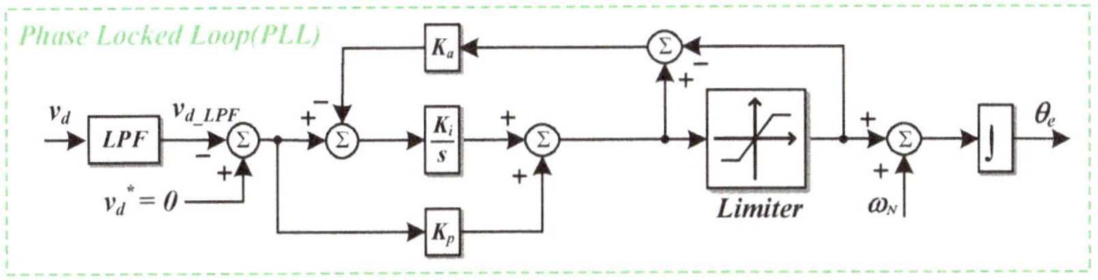

Figure 6. Control block diagram of phase detector using first order low pass filter and proportional integral (PI) controller.

3. Proposed Control Scheme of 3-Phase AC/DC PWM Converter Using SOGI-FLL

In the conventional PLL method that is discussed in Section 2.2, the sensed 3-phase input voltage is divided into magnitude and phase angle information by d-q transformation and the d-axis voltage in the synchronous reference frame is used to obtain phase angle information. However, the conventional PLL method is difficult to use in single-phase systems. In addition, if the cut-off frequency of the low pass filter is decreased to reduce the low frequency harmonics of the input power source, the total control bandwidth of the PLL will be reduced, which is disadvantageous to the sudden change of phase angle [17].

Therefore, this section describes the FLL method using SOGI, which is applicable to both single-phase and 3-phase systems and has excellent performance for low frequency harmonics reduction [18–20]. This section also introduces a method of detecting positive sequence voltage while using SOGI-FLL and it describes the design of the controller.

3.1. Second Order Generalized Integrator

Figure 7 shows the structure of the SOGI-based adaptive filter (AF) in a control block diagram [21]. The transfer function of the SOGI that is shown in Figure 7 can be expressed as Equation (6), and the transfer function for the output signals v' and qv' of the SOGI-based AF is defined as Equation (7). As can be seen from the transfer function $D(s)$ in Equation (7), the bandwidth of the SOGI-based AF is independent of the frequency ω' and it is only determined by the gain k. Also, the output signal qv' generates a signal that is delayed by 90 degrees from the phase of the output v', irrespective of the frequency of the center frequency ω' and the input signal v. Therefore, Equation (7) can be expressed as the bode plot as shown in Figure 8.

$$SG(s) = \frac{v'}{k\varepsilon_v}(s) = \frac{\omega' s}{s^2 + \omega'^2} \tag{6}$$

$$D(s) = \frac{v'}{v}(s) = \frac{k\omega' s}{s^2 + k\omega' s + \omega'^2}, \quad Q(s) = \frac{qv'}{v}(s) = \frac{k\omega'^2}{s^2 + k\omega' s + \omega'^2} \tag{7}$$

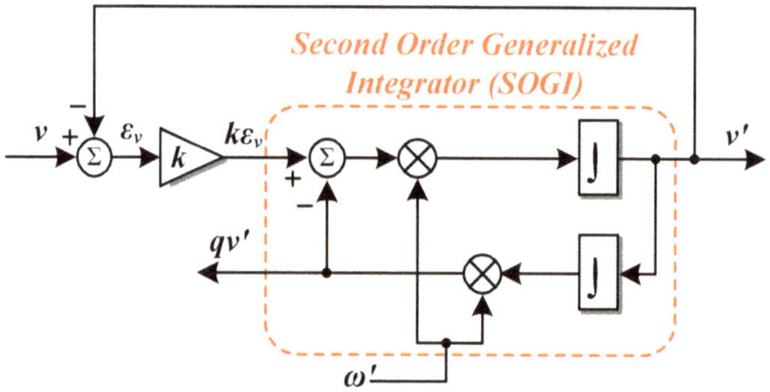

Figure 7. Control block diagram of second order generalized integrator (SOGI)-based adaptive filter (AF) (= SOGI-quadrature signal generator (QSG)).

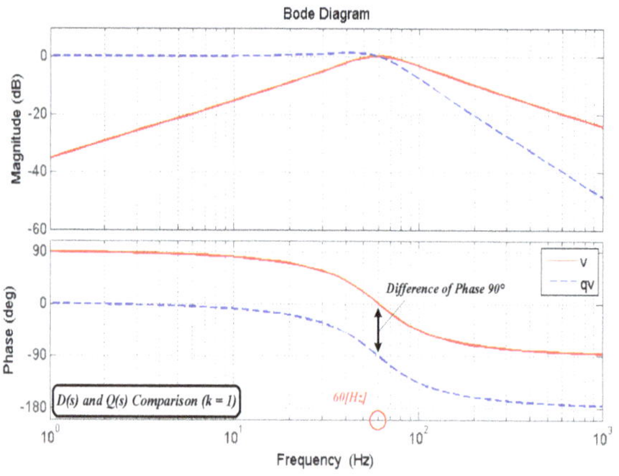

Figure 8. Comparison of transfer function $D(s)$ and $Q(s)$.

Figure 9 shows the bode plot of the transfer function of Equation (7). From the figure, it can be seen that as the gain k decreases, the filtering effect improves, but the stability according to the frequency change decreases. The SOGI-based AF, as shown in Figure 7, is also called the SOGI-quadrature signal generator (SOGI-QSG).

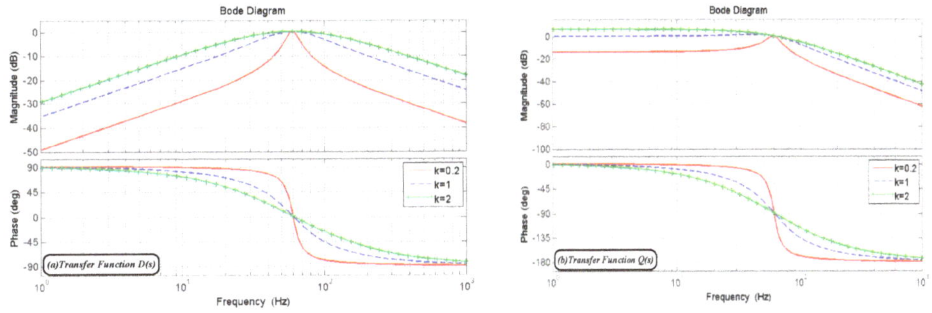

(**a**) Bode plot of transfer function $D(s)$ (**b**) Bode plot of transfer function $Q(s)$

Figure 9. Characteristics of SOGI-based AF various gain k.

3.2. SOGI-FLL

SOGI-FLL is used for detecting input voltage exactly, even if input frequency is varied and due to grid distortion or fault. Configuration and discretization of the SOGI-FLL technique in single-phase grid-connected inverter system is introduced in [22]. In this subsection, not only introducing the structure of SOGI-FLL, but also how SOGI-FLL extracts information of input voltage and tracks frequency variation based on its mathematical explanation and bode plot. If the center frequency ω' is set to the same frequency as the input signal, the SOGI-based AF outputs v' having the same amplitude and phase as the input signal and qv' delayed only by 90 degrees from the input signal. However, if the frequency of the input signal varies, the amplitudes of v' and qv' is changed. Therefore, it is not appropriate to use SOGI only for power conversion systems that use AC power such as grid-connected systems. In order to apply SOGI to a system where the input frequency is possibly changed, the center frequency must be able to follow the frequency change of the input signal.

In the FLL method, frequency information is extracted using the phase delayed signal of SOGI qv', which is used as the center frequency ω' of SOGI. FLL can be operated without the PI controller and trigonometric function calculations that were used in PLL. Figure 10 shows the control block diagram of SOGI-FLL, where ω_N is the angular frequency of the input voltage in steady state, as used in the PLL.

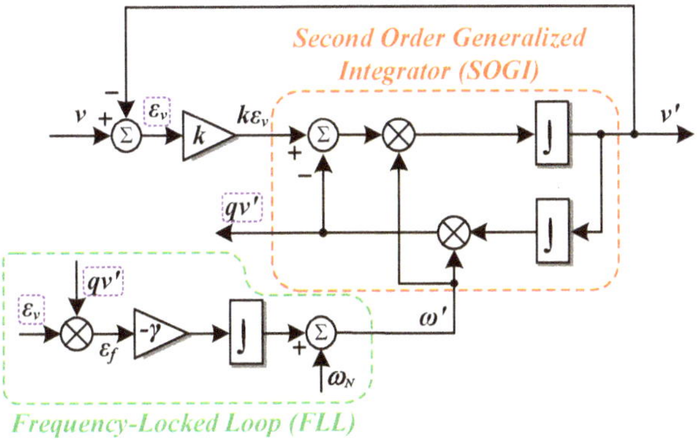

Figure 10. Control block diagram of SOGI-frequency locked loop (FLL) for phase synchronization.

In order to explain the SOGI-based AF with FLL, the error signal ε_v, which is the relation between the output signals qv', v', and v, should be analyzed. The transfer function for the input signal v and the error signal ε_v is shown in Equation (8).

$$E(s) = \frac{\varepsilon_v}{v}(s) = \frac{s^2 + \omega\prime^2}{s^2 + k\omega' s + \omega\prime^2} \tag{8}$$

The transfer function $E(s)$ is illustrated in the bode plot that is shown in Figure 11.

As shown in Figure 11, if the frequency of the input signal ω is lower than the center frequency ω', qv', and ε_v have the same phase. That is, the frequency error variable ε_f can be calculated using the two signals qv' and ε_v. If $\omega < \omega'$, then ε_f is a positive value, if $\omega = \omega'$ then $\varepsilon_f = 0$ and if $\omega > \omega'$ then ε_f is a negative value. This means that the negative gain $-\gamma$ and the integrator expressed in Figure 10 are used to set the DC component of the frequency error variable ε_f to zero so that the center frequency ω' tracks the frequency of the input signal ω.

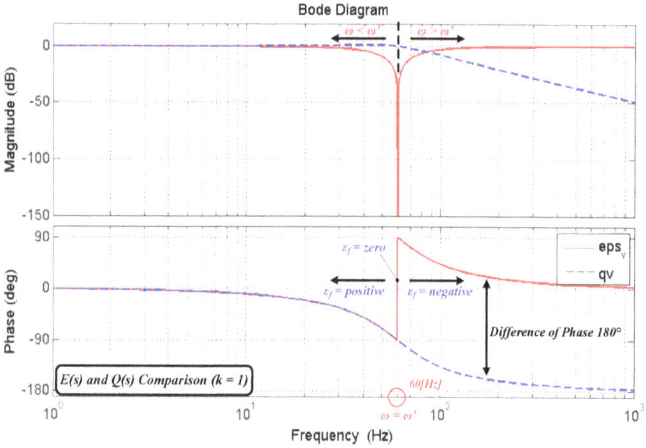

Figure 11. Comparison of transfer functions $E(s)$ and $Q(s)$.

To use this SOGI-FLL in 3-phase systems, two SOGIs are used. This is called a double second order generalized integrator (DSOGI). Figure 12 shows a control block diagram of the DSOGI-FLL that is used in the 3-phase AC system.

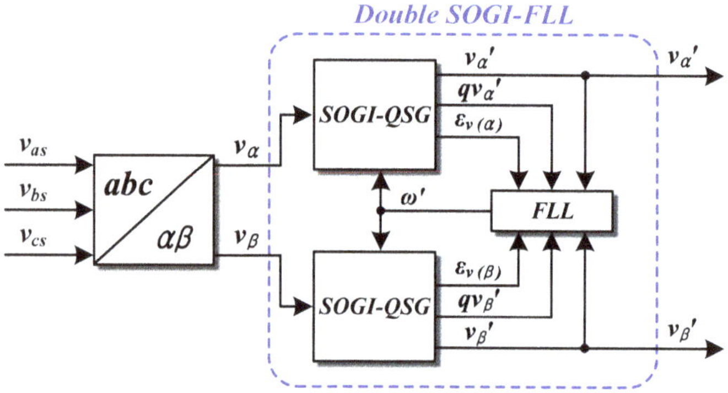

Figure 12. 3-phase SOGI-FLL (= DSOGI-FLL).

3.3. Positive Sequence Voltage Extraction and Phase Detection Using 3-Phase SOGI-FLL

In 1937, Lyon proposed a method for decomposing balanced 3-phase voltages v_a, v_b, and v_c into positive, negative, and zero sequence voltage in the time domain, respectively, while in 1918, Fortescue proposed a method in the frequency domain [23]. Using these methods, the balanced 3-phase voltage is decomposed into positive, negative, and zero sequence voltages, as shown in Equations (9)–(11), respectively.

$$\begin{bmatrix} v_a^+ \\ v_b^+ \\ v_c^+ \end{bmatrix} = \frac{1}{3}\cdot\begin{bmatrix} 1 & \alpha & \alpha^2 \\ \alpha^2 & 1 & \alpha \\ \alpha & \alpha^2 & 1 \end{bmatrix}\cdot\begin{bmatrix} v_a \\ v_b \\ v_c \end{bmatrix} \quad \left(v_{abc}^+ = [T_+]\cdot v_{abc}\right) \tag{9}$$

$$\begin{bmatrix} v_a^- \\ v_b^- \\ v_c^- \end{bmatrix} = \frac{1}{3}\cdot\begin{bmatrix} 1 & \alpha^2 & \alpha \\ \alpha & 1 & \alpha^2 \\ \alpha^2 & \alpha & 1 \end{bmatrix}\cdot\begin{bmatrix} v_a \\ v_b \\ v_c \end{bmatrix} \quad \left(v_{abc}^- = [T_-]\cdot v_{abc}\right) \tag{10}$$

$$\begin{bmatrix} v_a^0 \\ v_b^0 \\ v_c^0 \end{bmatrix} = \frac{1}{3}\cdot\begin{bmatrix} 1 & 1 & 1 \\ 1 & 1 & 1 \\ 1 & 1 & 1 \end{bmatrix}\cdot\begin{bmatrix} v_a \\ v_b \\ v_c \end{bmatrix} \quad \left(v_{abc}^0 = [T_0]\cdot v_{abc}\right) \tag{11}$$

where $\alpha = e^{j\frac{2\pi}{3}} = -\frac{1}{2} + j\frac{\sqrt{3}}{2}$ and $\alpha^2 = e^{j\frac{4\pi}{3}} = -\frac{1}{2} - j\frac{\sqrt{3}}{2}$.

3×3 matrices in each equation can be expressed as $[T_+]$, $[T_-]$, and $[T_0]$. If the balanced 3-phase voltage is transformed to the voltage in the stationary reference frame, it can be defined as Equation (12). Equation (12) is then simplified to Equation (13), with the assumption that no zero sequence components exist in the balanced 3-phase voltage.

$$\begin{bmatrix} v_\alpha \\ v_\beta \\ v_0 \end{bmatrix} = \frac{2}{3}\cdot\begin{bmatrix} 1 & -\frac{1}{2} & -\frac{1}{2} \\ 0 & \frac{\sqrt{3}}{2} & -\frac{\sqrt{3}}{2} \\ \frac{1}{2} & \frac{1}{2} & \frac{1}{2} \end{bmatrix}\cdot\begin{bmatrix} v_a \\ v_b \\ v_c \end{bmatrix} \quad v_{\alpha\beta 0} = \left[T_{\alpha\beta 0}\right]\cdot v_{abc} \tag{12}$$

$$\begin{bmatrix} v_\alpha \\ v_\beta \end{bmatrix} = \frac{2}{3}\cdot\begin{bmatrix} 1 & -\frac{1}{2} & -\frac{1}{2} \\ 0 & \frac{\sqrt{3}}{2} & -\frac{\sqrt{3}}{2} \end{bmatrix}\cdot\begin{bmatrix} v_a \\ v_b \\ v_c \end{bmatrix} \quad v_{\alpha\beta} = \left[T_{\alpha\beta}\right]\cdot v_{abc} \tag{13}$$

The transpose matrix $\left[T_{\alpha\beta}\right]$ is given by the following Equation (14).

$$\left[T_{\alpha\beta}\right]^T = \frac{3}{2}\cdot\begin{bmatrix} \frac{2}{3} & 0 \\ -\frac{1}{3} & \frac{\sqrt{3}}{3} \\ -\frac{1}{3} & -\frac{\sqrt{3}}{3} \end{bmatrix} \tag{14}$$

Using Equations (9)–(14), the positive and negative sequence voltages in the stationary reference frame can be summarized, as shown in Equations (15) and (16).

$$\begin{aligned} v_{\alpha\beta}^+ &= \left[T_{\alpha\beta}\right]\cdot v_{abc}^+ = \left[T_{\alpha\beta}\right]\cdot[T_+]\cdot v_{abc} \\ &= \left[T_{\alpha\beta}\right]\cdot[T_+]\cdot\left[T_{\alpha\beta}\right]^T\cdot v_{\alpha\beta} = \frac{1}{2}\cdot\begin{bmatrix} 1 & -q \\ q & 1 \end{bmatrix}\cdot v_{\alpha\beta} \end{aligned} \tag{15}$$

$$\begin{aligned} v_{\alpha\beta}^- &= \left[T_{\alpha\beta}\right]\cdot v_{abc}^- = \left[T_{\alpha\beta}\right]\cdot[T_-]\cdot v_{abc} \\ &= \left[T_{\alpha\beta}\right]\cdot[T_-]\cdot\left[T_{\alpha\beta}\right]^T\cdot v_{\alpha\beta} = \frac{1}{2}\cdot\begin{bmatrix} 1 & q \\ -q & 1 \end{bmatrix}\cdot v_{\alpha\beta} \end{aligned} \tag{16}$$

In Equations (15) and (16), $q = e^{-j\frac{\pi}{2}}$, which indicates that the phase is delayed by 90 degrees. In this case, the positive and negative sequence voltage extractors are added. Figure 12 is then modified, as shown in Figure 13.

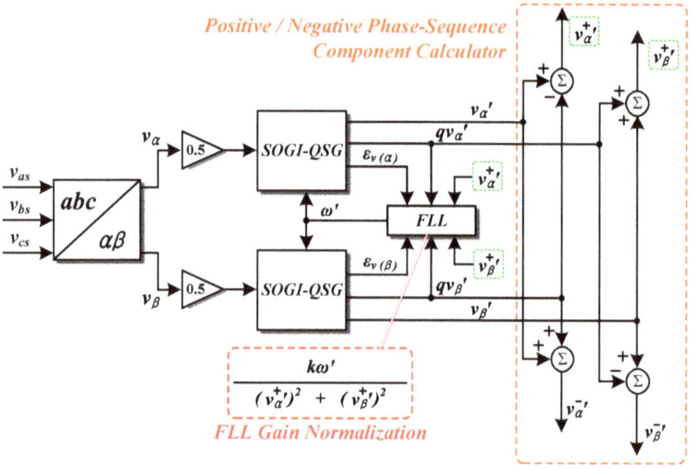

Figure 13. Positive sequence voltage extractor using 3-phase SOGI-FLL.

3.4. Control Block Diagram of Overall System with SOGI-FLL

Figure 14 shows the overall control block diagram of the proposed control scheme. Similar to the conventional control method, each line-to-line voltage of the grid is sensed and transformed to a stationary reference frame. The phase angle information is then extracted while using DSOGI-FLL. The coordinate of the input current of the 3-phase AC/DC PWM converter is transformed using the extracted phase angle derived from DSOGI-FLL. At this time, the output of the PI voltage controller that controls the dc-link voltage is applied to the input of the PI current controller that controls the active and reactive power of the grid. When the output of the current controller performs an inverse *d*-*q* transformation, an extracted phase angle using DSOGI-FLL is also used.

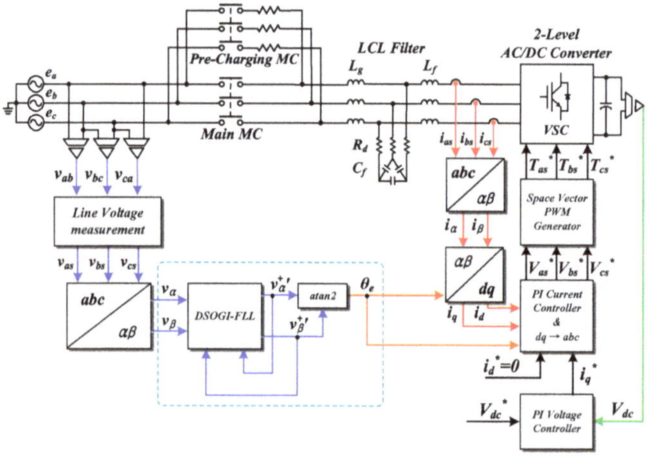

Figure 14. Control block diagram of 3-phase AC/DC PWM converter with SOGI-FLL.

As can be seen from the control block diagram, when using the proposed control scheme, the coordinate transformation is simpler than the control method while using the conventional PLL. In addition, the low pass filter that affects the bandwidth of the controller disappears in the proposed control method.

4. Simulation Results

In this paper, the simulation was performed before verifying the operation of the 3-phase AC/DC PWM converter with the SOGI-FLL method for the DC distributed system. Table 1 shows the parameters that are used in the simulation, while the configuration of the simulation circuit is shown in Figure 15. In the existing AC power grid, the power from the power plant is supplied to the consumer in single phase 220 V_{rms} and 3-phase 380 V_{rms}. Therefore, the 3-phase voltage is selected, as shown in the parameter. Also, when configuring of the DC distribution system, the DC-link voltage was selected as 700 V_{dc}, which is capable of meeting the scope of international electrotechnical commission (IEC) regulations [24]. The switching frequency is selected in consideration of the FS300R12KE3, which IGBT are used in this system, and the LCL filter are designed to satisfy the current distortion regulation of the IEEE ST 519-2014.

Table 1. Simulation parameter of DC distributed system.

Parameter	Value	Unit
Rated power of system	50	[kW]
3 phase line-to-line grid voltage	380	[V_{rms}]
Grid frequency	60	[Hz]
Switching frequency	5	[kHz]
Output DC-link voltage	700	[V_{dc}]
DC-link capacitance	10200	[μF]
LCL filter inductance at grid side	120	[μH]
LCL filter capacitance	50	[μF]
LCL filter inductance at converter side	500	[μH]

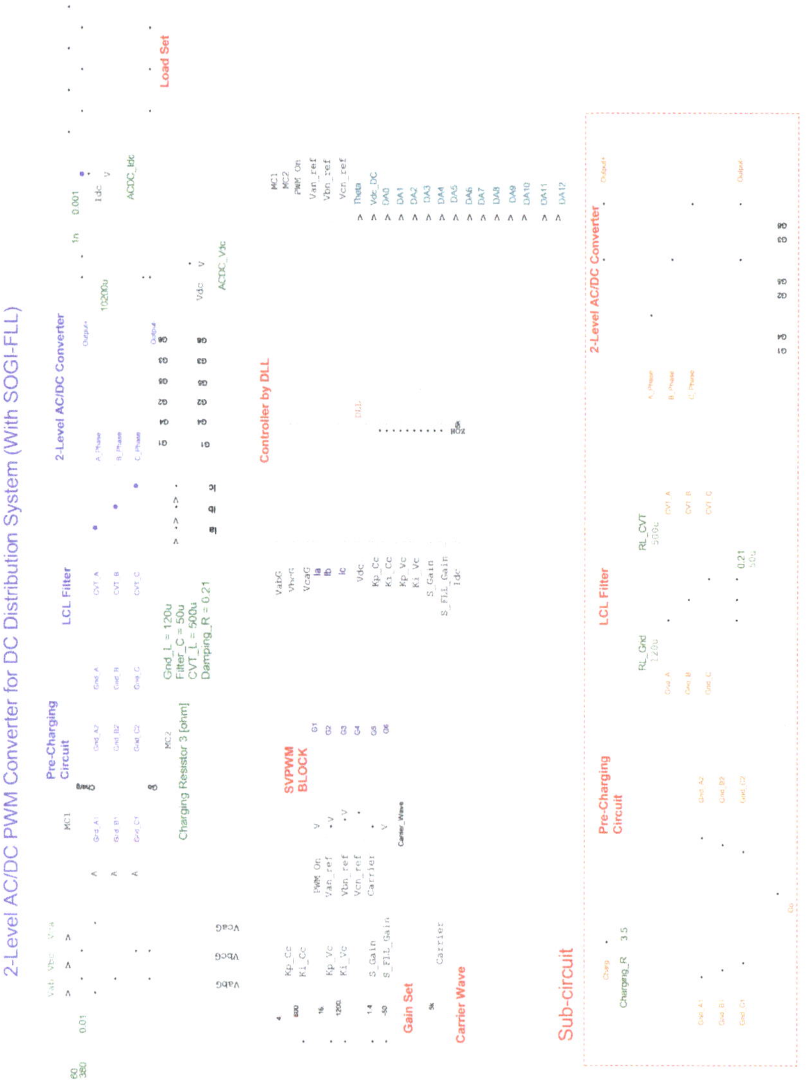

Figure 15. 3-phase AD/DC PWM converter schematic of DC distribution system used in simulation.

Figure 16a shows the output voltage waveform of a DC distribution system while using a 3-phase AC/DC PWM converter. The system uses an initial charge circuit that charges a certain amount of voltage to the output capacitor for 0.2 s to reduce the inrush current entering the output capacitor. Then, the charge of the capacitor voltage increases, while the initial charge circuit is changed from the initial charge to a standby condition. Finally, an output voltage control of 700 V_{dc} is initiated from 0.3 s. In order to prevent the high overshoot of the output voltage at the start of the control, the voltage reference is given in the form of a ramp. The voltage reference then rapidly enters the steady state due to the fast dynamic characteristics of the PI controller.

Figure 16b shows the output current waveform after connecting the DC link output to the load. The load is connected to the system from 0.8 s and the load capacity is increased by 10 kW every 0.15 s. Finally, a current is stably generated at the system rated capacity of 50 kW.

Figure 16c shows the 3-phase input current waveform from the grid to the system. It is confirmed that the reactive component rarely flows when no load is applied, and the magnitude of the input current increases step by step when the load is connected.

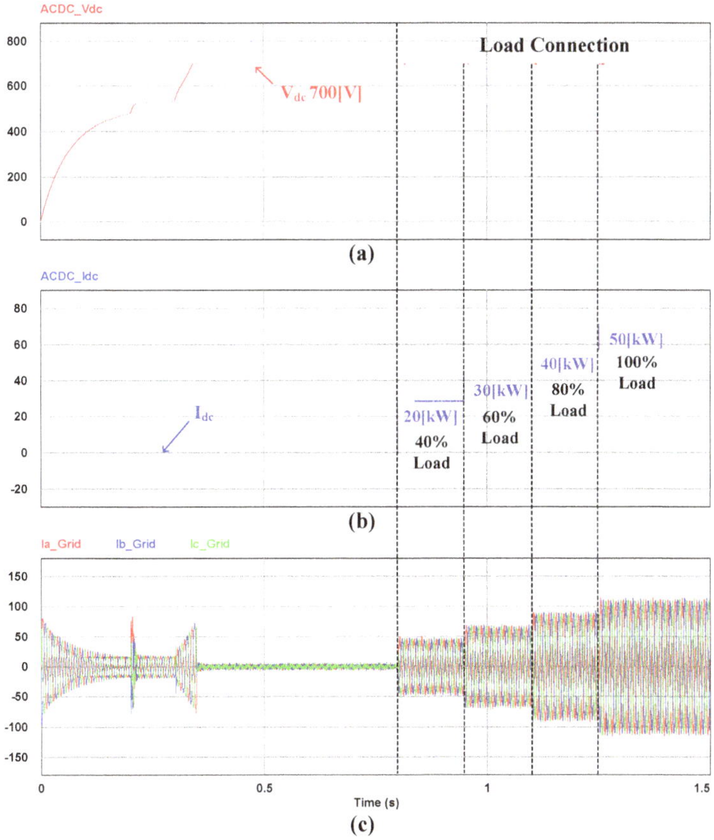

Figure 16. Input and output waveforms of DC distribution system (**a**) the dc-link voltage; (**b**) the current of the DC load; and, (**c**) the current of the grid.

Figure 17a shows the 3-phase input voltage waveforms containing the 3rd, 5th, 7th, 9th, and 11th harmonics. The input voltage is detected through the sensor and is then input to the controller. The SOGI-FLL is applied to the system controller in the middle of the $\alpha\beta - dq$ coordinate transformation, and the output voltage of the SOGI-FLL indicates that the harmonics are reduced, as shown in Figure 17b. The THD of the detected positive sequence voltage is about 3.9%, which is appropriate for the phase angle detection.

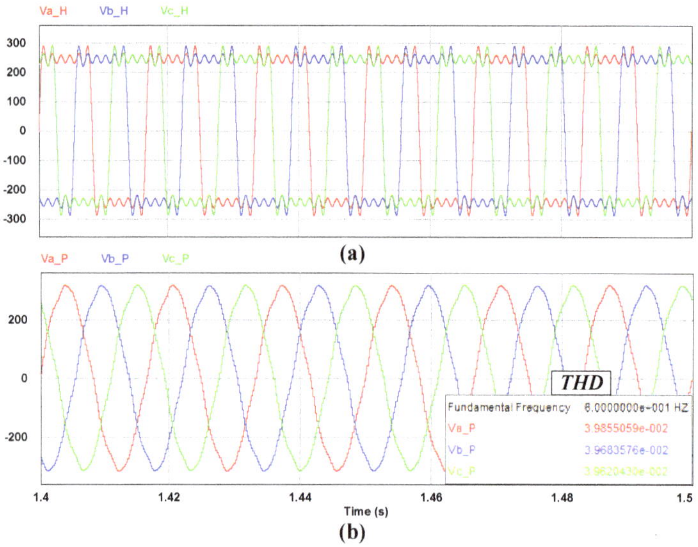

Figure 17. Waveform of voltage when harmonics is included in input side; (**a**) the 3-phase input voltage with harmonic components; and, (**b**) the 3-phase input voltage detected by the SOGI-FLL.

Figure 18 shows the waveforms when a voltage drop of about 30% occurs on the *b* phase. Figure 18a shows the dc-link output voltage waveform. A voltage ripple of about 2 [$V_{peak\text{-}to\text{-}peak}$] is generated while voltage drop occurs. Figure 18b shows the unbalanced 3-phase input voltage waveform. The balanced 3-phase condition is distorted by the voltage drop from 1 to 1.2 s. Figure 18c shows the detected positive sequence voltage waveform using SOGI-FLL. Unlike that shown in Figure 18b, the magnitude of voltage is slightly reduced and it maintains a balanced 3-phase condition. The 3-phase input current waveform is illustrated in Figure 18d. Unlike the voltage in Figure 18b, the changes in the amplitudes of the phases differ, and the shape of the waveform is severely distorted. Figure 18e shows the center frequency ω' extracted from SOGI-FLL. Only a slight ripple is generated, and it does not significantly differ from the frequency value in the steady state.

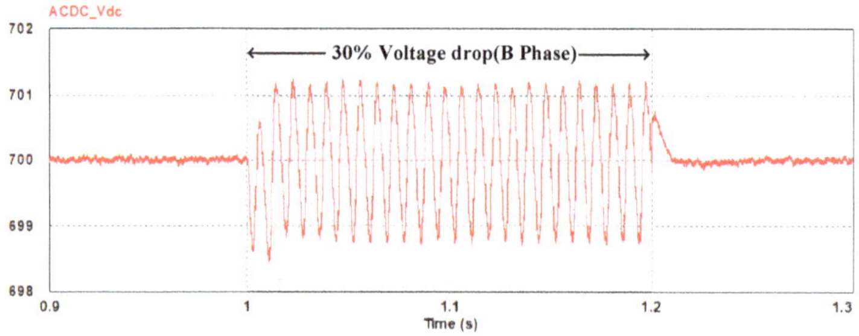

(a)

Figure 18. *Cont.*

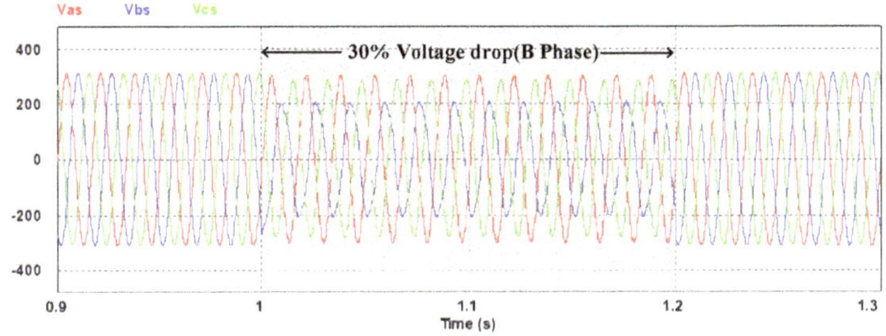

(b)

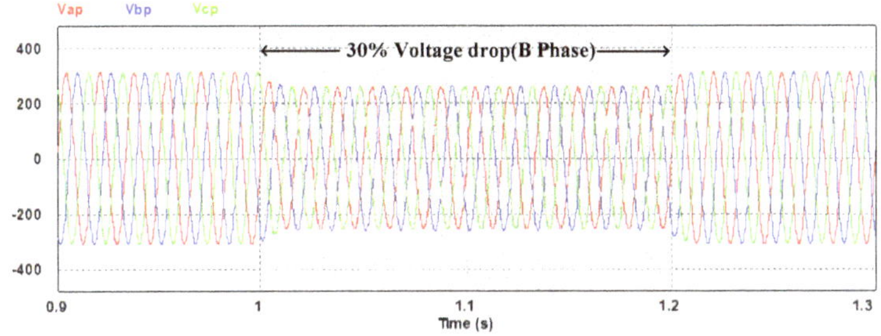

(c)

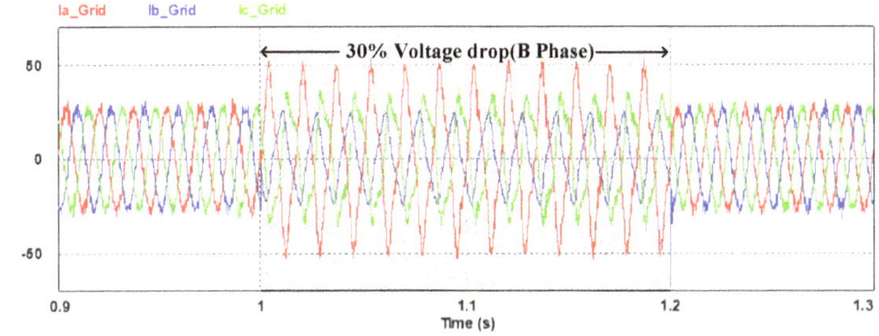

(d)

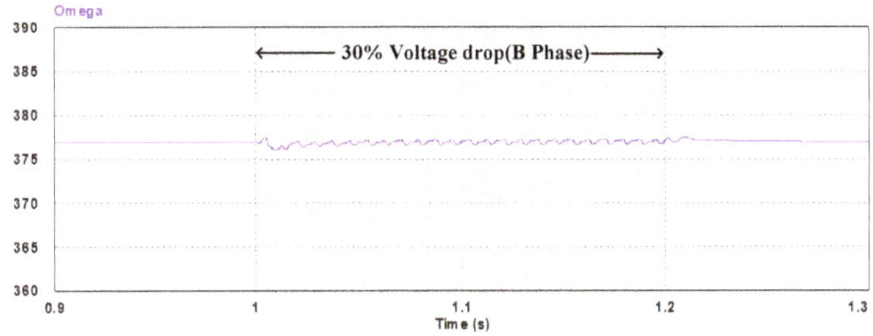

(e)

Figure 18. Waveforms of simulation results when *b* phase has a voltage drop; (**a**) the dc-link voltage; (**b**) the 3-phase input voltage with voltage unbalance; (**c**) the 3-phase input positive voltage detected by SOGI-FLL; (**d**) the 3-phase input current; and, (**e**) the center frequency extracted from SOGI-FLL.

Figure 19 shows the positive and negative sequence voltages and the phase angle generated by SOGI-FLL. While only negligible negative sequence voltage occurs before the unbalance condition, the negative sequence voltage increases immediately after the unbalance condition. In addition, while the phase angle of the positive sequence voltage is stably tracked, the phase angle of the negative sequence component is rapidly changed.

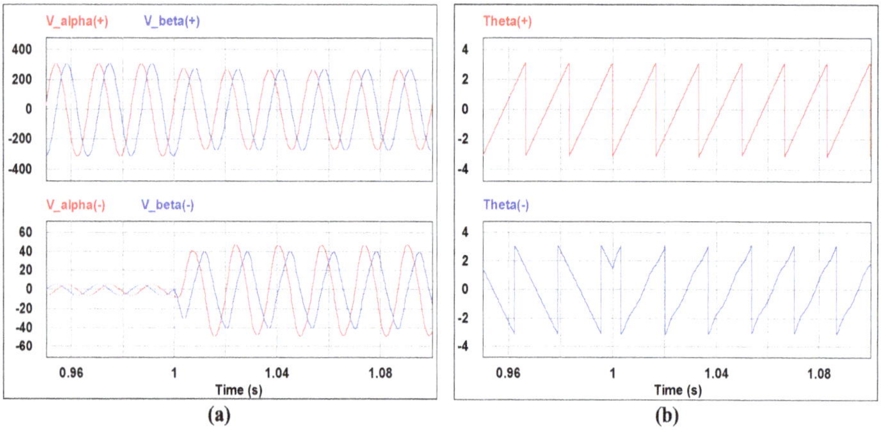

Figure 19. Waveforms of simulation results when voltage drop is occurred; (**a**) the positive and negative sequence voltage; and, (**b**) the phase angle of the positive and negative sequence voltage.

Figure 20 shows the waveforms when the grid frequency decreased from 60 Hz to 45 Hz. Grid frequency is reduced to 45 Hz from 1 to 1.2 s, while the power is supplied to the 12 kW load, and the grid frequency is then recovered to 60 Hz. Figure 20a shows that the dc-link output voltage waveform maintains an almost constant voltage, even when the frequency decreases. Figure 20b shows the 3-phase input voltage waveform. When the frequency is decreased from 1 to 1.2 s, the period increases. Figure 20c shows the positive sequence voltage waveform that was detected by SOGI-FLL. As shown in Figure 20b, since only negligible distortion occurs in the voltage, only the period changes without large amplitude change. Figure 20d shows the 3-phase input current waveform. It can be seen that, while the input current also only changes the period, it but does not cause large distortion. Figure 20e shows the center frequency ω', as extracted from SOGI-FLL. As soon as the grid frequency changes, the FLL follows the grid frequency and synchronizes the output center frequency to the grid frequency. The FLL gain value is designed considering the frequency tracking time of 0.1 s, but the steady state takes slightly more time than the designed time due to the change of the grid conditions.

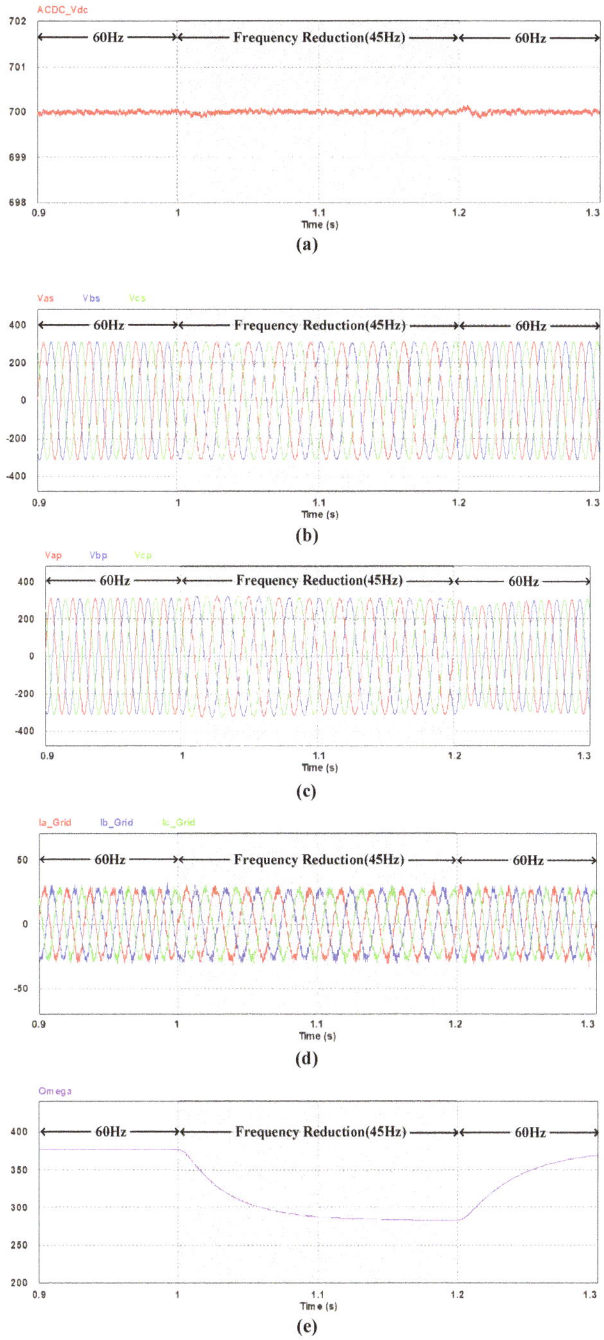

Figure 20. Waveforms of simulation results when grid frequency decreased from 60 Hz to 45 Hz drop; (**a**) the dc-link voltage; (**b**) the 3-phase input voltage; (**c**) the 3-phase input positive voltage detected by SOGI-FLL; (**d**) the 3-phase input current; and, (**e**) the center frequency that was extracted from SOGI-FLL.

In Figure 21, the conventional synchronous reference frame-phase locked loop (SRF-PLL) method [25] and SOGI-FLL method are compared to each other when the 3-phase grid voltage frequency is decreased from 60 Hz to 45 Hz. During power supply to 50 kW load, the grid frequency decreases to 45 Hz from 1 to 1.2 s, and then recovers to 60 Hz.

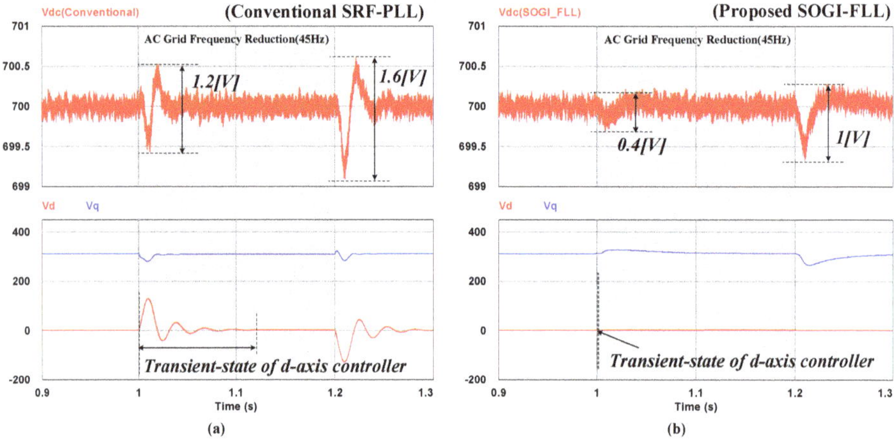

Figure 21. Waveforms of simulation results when frequency suddenly change (**a**) the DC-link voltage, *d* and *q*-axis voltage in synchronous reference frame with conventional synchronous reference frame-phase locked loop (SRF-PLL) (**b**) the DC-link voltage, *d* and *q*-axis voltage in synchronous reference frame with proposed control method.

Figure 21a shows that the *d*-axis voltage in a synchronous reference frame oscillates significantly when the grid frequency changes, and the dc-link voltage oscillation ranges from ±0.5 to 0.9 V_{dc}. This oscillation of dc-link output voltage has the potential to adversely affect both the load and the dc-link capacitor. The oscillation occurs because the bandwidth of the controller is low due to the low cut-off frequency of the low pass filter. While this problem can be solved by increasing the cut-off frequency, the low frequency harmonics cannot be blocked.

On the other hand, in the voltages in the synchronous reference frame using SOGI-FLL, as shown in Figure 20b, no oscillation occurs in the *d*-axis voltage during the frequency change, and the transient state of the dc-link voltage is improved by about 50%.

As a result, the conventional SRF-PLL method takes a long time to stabilize the overshoot of the *d*-axis current controller in case of a frequency change because of the bandwidth problems in the controller due to digital filters, however, the proposed control method that is based on SOGI-FLL does not use these digital filters, so it is possible to stable and fast transient control.

5. Experiment Results

An experiment was performed to verify the feasibility of the proposed control method that is applied in a 3-phase AC/DC PWM converter for a DC distribution system. The configurations of the experimental system and power stacks are shown in Figures 22–24, and the experimental parameters are shown in Table 1.

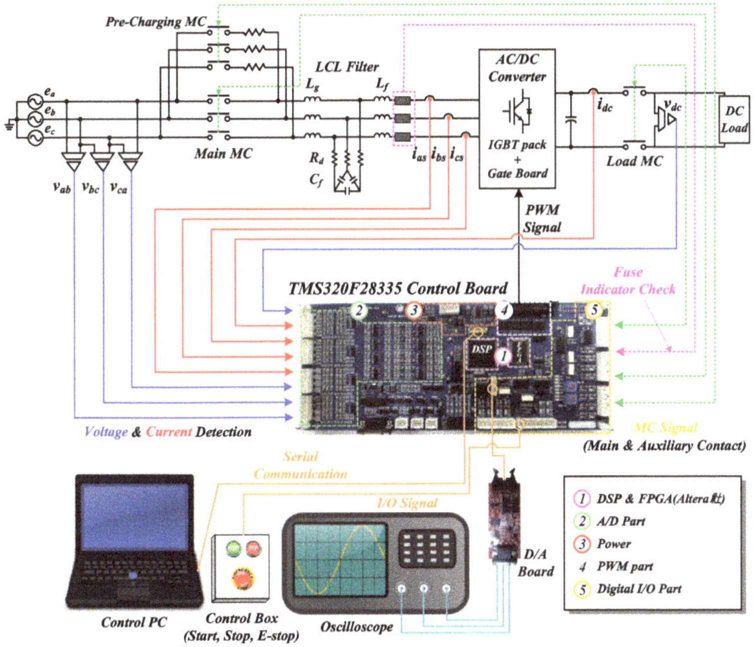

Figure 22. Configuration of the experiment system. MC: Magnetic Contactor.

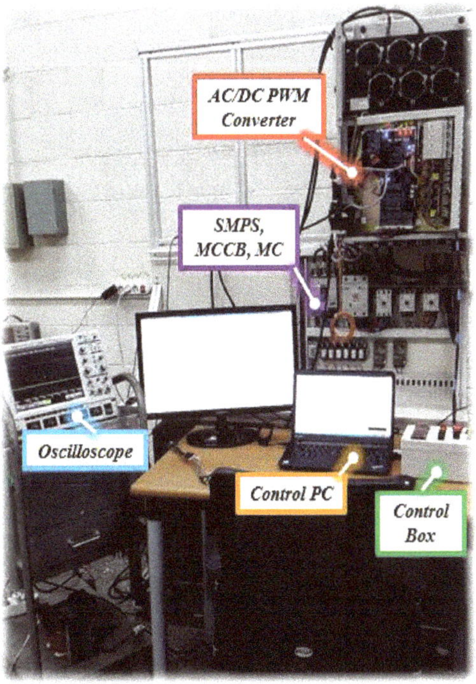

Figure 23. Experiment setup 1 of the 3-phase AC/DC PWM converter.

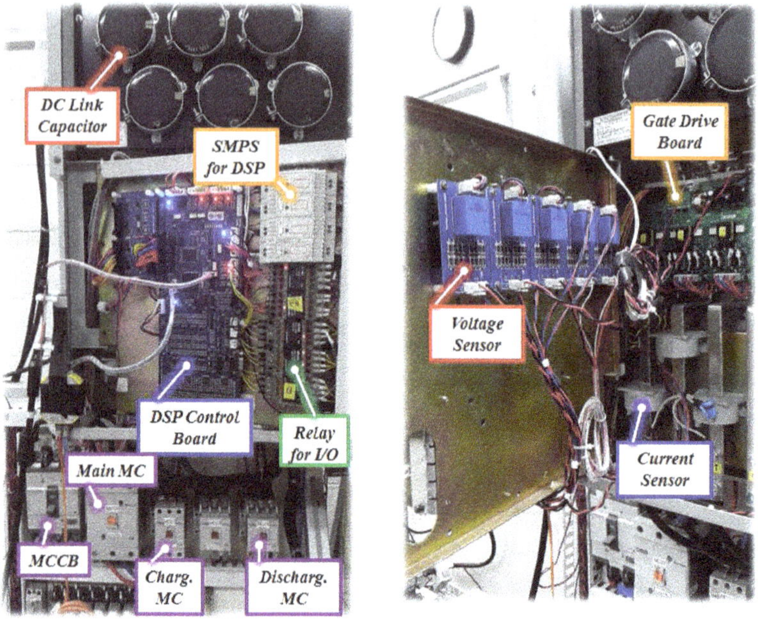

Figure 24. Experiment setup 2 of the 3-phase AC/DC PWM converter. SMPS: switched-mode power supply; DSP: digital signal processor; MCCB: molded case circuit breaker.

Figure 25 shows the waveforms for the operating sequence of a 3-phase AC/DC PWM converter for DC distribution. To prevent an excessive inrush current to the capacitor, the dc-link capacitor is charged while using an initial charging circuit consisting of magnetic switch and a resistor. After the initial charging is completed, the main magnetic switch is activated and the voltage is charged to the capacitor up to the input phase voltage peak value of 311 V_{dc}. The controller then starts to operate and the voltage increases up to 700 V_{dc}, which is the voltage reference, and the operation maintains steady state. As the load capacity increases, the output current gradually increases and the voltage remains constant, despite the increasing load.

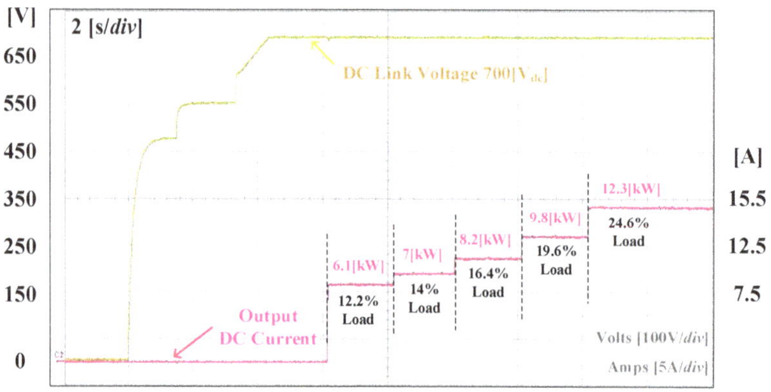

Figure 25. Waveform of 3-phase AC/DC PWM converter operation.

Figure 26 shows the input phase current waveform at the grid side when a load of 12.3 kW is applied, which is the largest capacity used of those shown in Figure 25. Since the load capacity is lower than that of the designed capacity of 50 kW, the current waveform includes the harmonics.

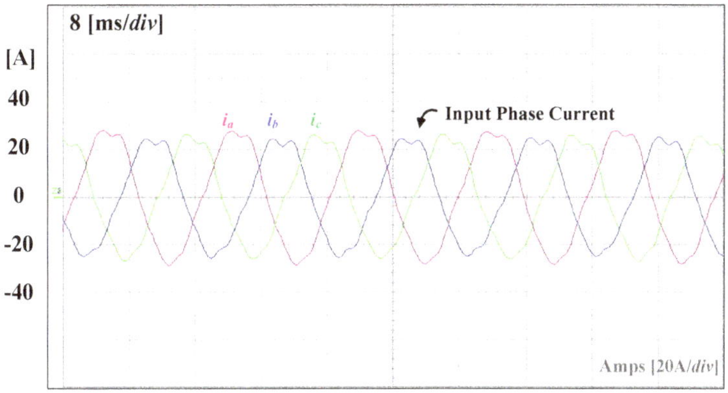

Figure 26. Waveform of input phase current of 3-phase AC/DC PWM converter.

Figure 27 shows the grid side input a voltage and current waveforms. The phase angle of the input phase voltage was extracted while using the positive sequence voltage and phase detector using SOGI-FLL without the need to use the PLL of the conventional control or the PLL used for the coordinate transformation of the phase current. In addition, the power factor is controlled to be 1 by adjusting the d-axis current command in the synchronous reference frame, which is a reactive power component. The phase detection performance of SOGI-FLL is then verified by experiment waveform.

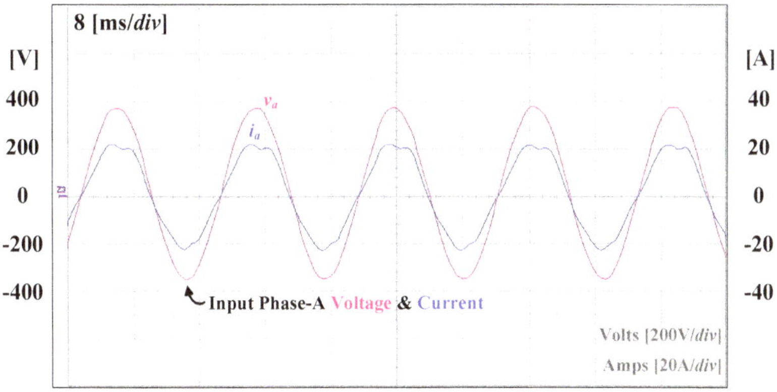

Figure 27. Waveform of input phase a voltage and current.

Figure 28 shows the 3-phase voltage waveform when a voltage drop (about 10%) occurs on the b phase. Figure 28b shows the expanded waveform of Figure 28a when the voltage drop occurs, and Figure 28c shows the waveform of the 3-phase positive sequence input voltage extracted while using SOGI-FLL. The amplitude of the 3-phase positive sequence voltage decreases due to the voltage drop. Even though the input 3-phase voltage is unbalanced, the extracted positive sequence voltage using SOGI-FLL maintains the balanced condition.

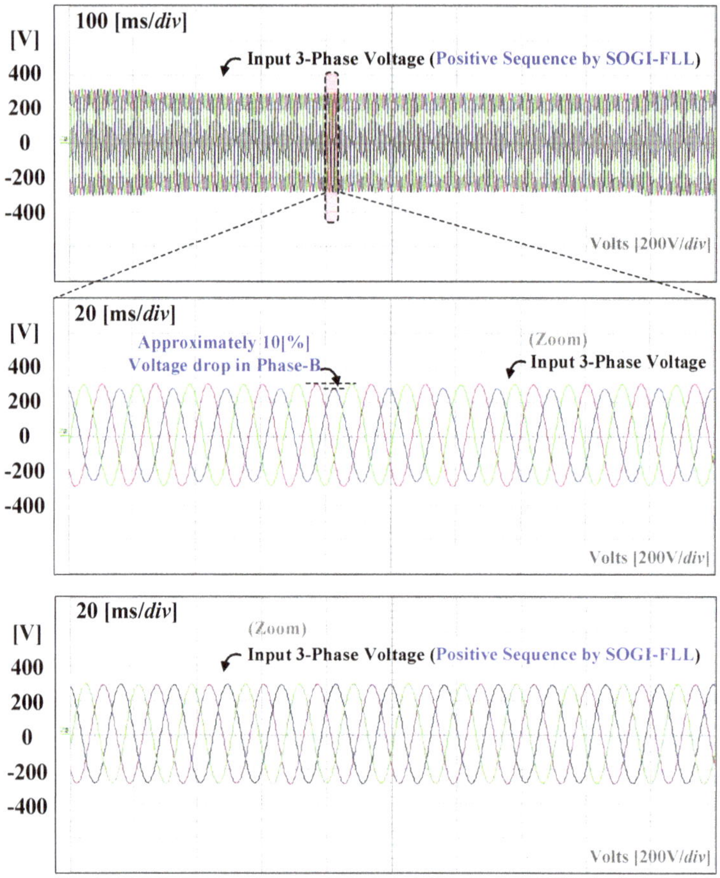

Figure 28. Waveforms of extracted 3-phase positive sequence input voltage using SOGI-FLL when the b phase has voltage drop.

Figure 29 shows the positive and negative sequence $\alpha\beta$ voltage and phase angle θ, before and after the b phase voltage drop. In the case of the $\alpha\beta$ voltage in the stationary reference frame, the amplitude and phase angle of the positive sequence component are always the same as these of the grid side. While the appearance of the negative sequence component is negligible at the steady state, the amplitude varies depending on the distortion in the unbalance condition. The phase angle θ of the positive sequence is always rotated in the same direction. In the case of a negative sequence, the magnitude, direction, and shape of the phase angle change depending on the balance, unbalance, and distortion.

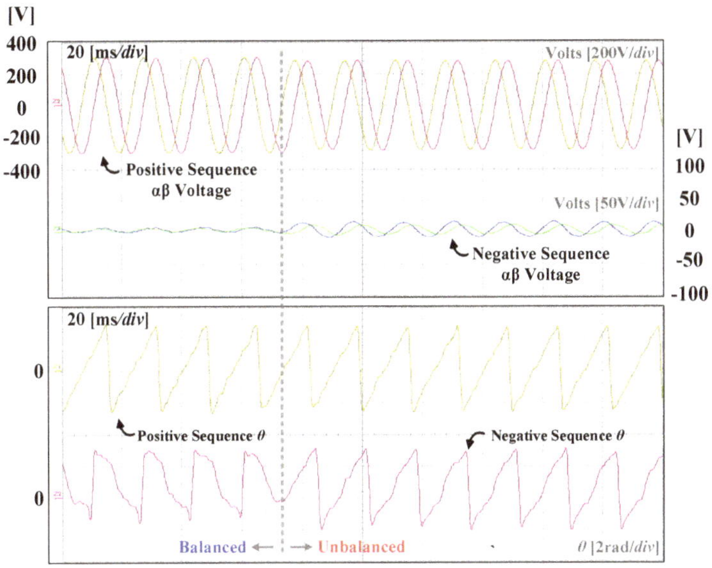

Figure 29. Waveform of detected positive/negative sequence $\alpha\beta$ voltage and phase angle θ (before and after b phase voltage drop).

6. Conclusions

In this paper, we studied a control method using SOGI-FLL for a 3-phase AC/DC PWM converter that is used in a DC distribution system. The proposed control method comprises a SOGI-FLL, which performs positive sequence voltage extraction, phase angle extraction, and harmonic filtering without an additional filter.

In DC distribution system, the power delivered to the load will depend on the 3-phase AC/DC PWM converter that is connected to the grid. However, the conventional control method with PLL used in 3-phase AC/DC PWM converter can not flexibly cope with the various conditions occurred in the grid. Unlike the conventional method, the proposed control method using SOGI-FLL can allow for the stable power to supply to the load in DC distribution even when various conditions occurred in the grid because it have stable and fast transient characteristics.

Experiment and simulation applied to 50 kW 3-phase 2-level AC/DC PWM converter was performed to verify the feasibility and effectiveness of the proposed control method. As a result, stable and fast transient control was verified while using the proposed method, even if the input voltage includes various harmonics, the frequency changes and a voltage drop occurs. Given those results, it is expected that the proposed method based on SOGI-FLL is one of good candidates for DC distribution system.

Author Contributions: J.-W.K., and K.-W.S. conceived and designed the experiment; J.-W.K. and K.-W.S. performed the experiment; J.-W.K., K.-M.K., H.L. and K.-W.S. analyzed the theory. J.-W.K. and H.L. wrote the manuscript. J.K. and C.-Y.W. participated in research plan development and revised the manuscript. All authors have contributed to the manuscript.

Funding: This research received no external funding.

Acknowledgments: This work was supported by the Korea Institute of Energy Technology Evaluation and Planning (KETEP) and the Ministry of Trade, Industry & Energy (MOTIE) of the Republic of Korea. (No. 20162010103830). This work was supported by "Human Resources Program in Energy Technology" of the Korea Institute of Energy Technology Evaluation and Planning (KETEP), granted financial resource from the Ministry of Trade, Industry & Energy, Republic of Korea. (No. 20184030202190)

Conflicts of Interest: The authors declare no conflict of interest.

References

1. Kim, H.S.; Ryu, M.H.; Baek, J.W.; Jung, J.H. High-Efficiency Isolated Bidirectional AC–DC Converter for a DC Distribution System. *IEEE Trans. Power Electron.* **2013**, *28*, 1642–1654. [CrossRef]
2. Bosich, D.; Mastromauro, R.A.; Sulligoi, G. AC-DC interface converters for MW-scale MVDC distribution systems: A survey. In Proceedings of the 2017 IEEE Electric Ship Technologies Symposium (ESTS), Arlington, VA, USA, 15–17 August 2017; pp. 44–49.
3. Rodríguez, P.; Teodorescu, R.; Candela, I.; Timbus, A.V.; Liserre, M.; Blaabjerg, F. New positive-sequence voltage detector for grid synchronization of power converters under faulty grid conditions. In Proceedings of the 37th IEEE Power Electronics Specialists Conference, Jeju, Korea, 18–22 June 2006; pp. 1–7.
4. Bacon, V.D.; Oliveira da Silva, S.A. Performance improvement of a three-phase phase-locked-loop algorithm under utility voltage disturbances using non-autonomous adaptive filters. *IET Power Electron.* **2015**, *8*, 2237–2250. [CrossRef]
5. Karimi-Ghartemani, M.; Iravani, M.R. A method for synchronization of power electronic converters in polluted and variable-frequency environments. *IEEE Trans. Power Syst.* **2004**, *19*, 1263–1270. [CrossRef]
6. Jung, J.; Kim, H.; Ryu, M.; Kim, J.; Baek, J. Single-phase bidirectional AC-DC boost rectifier for DC distribution system. In Proceedings of the IEEE ECCE Asia Downunder, Melbourne, Australia, 3–6 June 2013; pp. 544–549.
7. Jung, T.; Gwon, G.; Kim, C.; Han, J.; Oh, Y.; Noh, C. Voltage Regulation Method for Voltage Drop Compensation and Unbalance Reduction in Bipolar Low-Voltage DC Distribution System. *IEEE Trans. Power Deliv.* **2018**, *33*, 141–149. [CrossRef]
8. Kim, H.; Ryu, M.; Baek, J.; Jung, J. High-Efficiency Isolated Bidirectional AC–DC Converter for a DC Distribution System. *IEEE Trans. Power Electron.* **2013**, *28*, 1642–1654. [CrossRef]
9. Rodríguez, P.; Luna, A.; Muñoz-Aguilar, R.S.; Etxeberria-Otadui, I.; Teodorescu, R.; Blaabjerg, F. A Stationary Reference Frame Grid Synchronization System for Three-Phase Grid-Connected Power Converters under Adverse Grid Conditions. *IEEE Trans. Power Electron.* **2012**, *27*, 99–112. [CrossRef]
10. Kang, J.; Hyun, S.; Hong, S.; Won, C. Advanced control method of 3-phase AC/DC PWM converter for DC distribution using the SOGI-FLL. In Proceedings of the IEEE 8th International Power Electronics and Motion Control Conference (IPEMC-ECCE Asia), Hefei, China, 22–26 May 2016; pp. 2120–2124.
11. Kang, J.W.; Lee, H.; Hyun, S.W.; Kim, J.; Won, C.Y. An Enhanced Control Scheme Based on New Adaptive Filters for Cascaded NPC/H-Bridge System. *Energies* **2018**, *11*, 1034. [CrossRef]
12. Jung, C.H. Control Method for Reducing Circulating Current in Parallel Operation of DC Distribution System for Building Applications. Master's Thesis, Sungkyunkwan University, Suwon, Korea, February 2013.
13. Han, D.W. Integrated Parallel Operation of Power Conversion Module for DC Distribution System. Master's Thesis, Sungkyunkwan University, Suwon, Korea, February 2014.
14. Park, Y.W. Parallel Control Using Current Droop Control of Cubic Equation for DC-grid Distribution Systems. Master's Thesis, Sungkyunkwan University, Suwon, Korea, August 2014.
15. Hadjidemetriou, L.; Kyriakides, E.; Blaabjerg, F. A New Hybrid PLL for Interconnecting Renewable Energy Systems to the Grid. *IEEE Trans. Ind. Appl.* **2013**, *49*, 2709–2719. [CrossRef]
16. Rodriguez, P.; Luna, A.; Ciobotaru, M.; Teodorescu, R.; Blaabjerg, F. Advanced Grid Synchronization System for Power Converters under Unbalanced and Distorted Operating Conditions. In Proceedings of the IECON 2006-32nd Annual Conference on IEEE Industrial Electronics, Paris, France, 6–10 November 2006; pp. 5173–5178.
17. Mnider, A.M.; Atkinson, D.J.; Dahidah, M.; Zbede, Y.B.; Armstrong, M. A programmable cascaded LPF based PLL scheme for single-phase grid-connected inverters. In Proceedings of the 7th International Renewable Energy Congress (IREC), Hammamet, Tunisia, 22–24 March 2016; pp. 1–6.
18. Yuan, X.M.; Merk, W.; Stemmler, H.; Allmeling, J. Stationary-frame generalized integrators for current control of active power filters with zero steady-state error for current harmonics of concern under unbalanced and distorted operating conditions. *IEEE Trans. Ind. Appl.* **2002**, *38*, 523–532. [CrossRef]
19. Rodriguez, P.; Luna, A.; Candela, I.; Teodorescu, R.; Blaabjerg, F. Grid synchronization of power converters using multiple second order generalized integrators. In Proceedings of the 34th Annual Conference of IEEE Industrial Electronics, Orlando, FL, USA, 10–13 November 2008; pp. 755–760.

20. Rodriguez, P.; Luna, A.; Etxeberria, I.; Hermoso, J.R.; Teodorescu, R. Multiple. Second order generalized integrators for harmonic synchronization of power converters. In Proceedings of the IEEE Energy Conversion Congress and Exposition, San Jose, CA, USA, 20–24 September 2009; pp. 2239–2246.
21. Xin, Z.; Qin, Z.; Lu, M.; Loh, P.C.; Blaabjerg, F. A new second-order generalized integrator based quadrature signal generator with enhanced performance. In Proceedings of the IEEE Energy Conversion Congress and Exposition (ECCE), Milwaukee, WI, USA, 18–22 September 2016; pp. 1–7.
22. Ciobotaru, M.; Teodorescu, R.; Blaabjergc, F. A new single-phase PLL structure based on second order generalized integrator. In Proceedings of the 37th IEEE Power Electronics Specialists Conference, Jeju, Korea, 18–22 June 2006; pp. 1–6.
23. Rocha, C.X.; Camacho, J.R.; Viajante, G.P. Fault detection in a three-phase system grid connected using SOGI structure to calculate vector components. In Proceedings of the International Conference on Renewable Energies and Power Quality (2015 ICREPQ), La Coruña, Spain, 25–27 March 2015; pp. 25–27.
24. Nam, K.Y.; Choi, S.B.; Ryoo, H.S.; Jeong, S.H.; Lee, J.D.; Jung, D.H.; Kim, D.K. Establishment of Test Field for Application of IEC 60364-4-44 in Korea. In Proceedings of the IEEE PES Power Systems Conference and Exposition, Atlanta, GA, USA, 29 October–1 November 2006; pp. 1944–1949.
25. Golestan, S.; Guerrero, J.M. Conventional Synchronous Reference Frame Phase-Locked Loop is an Adaptive Complex Filter. *IEEE Trans. Ind. Electron.* **2015**, *62*, 1679–1682. [CrossRef]

applied sciences

MDPI

Article

A Frequency–Power Droop Coefficient Determination Method of Mixed Line-Commutated and Voltage-Sourced Converter Multi-Infeed, High-Voltage, Direct Current Systems: An Actual Case Study in Korea

Gyusub Lee [1], Seungil Moon [1] and Pyeongik Hwang [2,*]

[1] Department of Electrical and Computer Engineering, Seoul National University, Seoul 08826, Korea; lgs1106@snu.ac.kr (G.L.); moonsi@snu.ac.kr (S.M.)
[2] Department of Electrical Engineering, Chosun University, Gwangju 61452, Korea
* Correspondence: hpi@chosun.ac.kr; Tel.: +82-62-230-7033

Received: 10 December 2018; Accepted: 9 February 2019; Published: 12 February 2019

Abstract: Among the grid service applications of high-voltage direct current (HVDC) systems, frequency–power droop control for islanded networks is one of the most widely used schemes. In this paper, a new frequency-power droop coefficient determination method for a mixed line-commutated converter (LCC) and voltage-sourced converter (VSC)-based multi-infeed HVDC (MIDC) system is proposed. The proposed method is designed for the minimization of power loss. An interior-point method is used as an optimization algorithm to implement the proposed scheduling method, and the droop coefficients of the HVDCs are determined graphically using the Monte Carlo sampling method. Two test systems—the modified Institute of Electrical and Electronics Engineers (IEEE) 14-bus system and an actual Jeju Island network in Korea—were utilized for MATLAB simulation case studies, to demonstrate that the proposed method is effective for reducing power system loss during frequency control.

Keywords: grid service of HVDC; frequency droop control; multi-infeed HVDC system; LCC HVDC; VSC HVDC; loss minimization

---

1. Introduction

High-voltage direct current (HVDC) systems have played an important role in sub-marine power transmission, due to economic advantages over their alternating current (AC) counterparts [1]. Nowadays, due to the complexity of modern power systems, HVDC systems are adopted in practice, not only for constant power delivery between massive networks, but also to provide grid services for island grids [2–8]. For example, in Korea, two line-commutated converter (LCC) HVDCs, the 180-kV/300-MW Haenam–Jeju HVDC [9] and ±250 kV/400 MW Jindo–Jeju HVDC [10], were constructed by the Korea Electric Power Corporation (KEPCO) for supporting Jeju Island and integrating massive wind power plants. Furthermore, KEPCO is planning to construct one more ±150 kV/200 MW voltage-sourced converter (VSC) HVDC between the main grid and Jeju Island [11].

With the technical development of both the LCC and VSC HVDC systems, there are numerous island networks with multi-infeed HVDC (MIDC) systems, encompassing both types of HVDCs, similar to the Jeju Island case. According to trends, many studies have concentrated on the MIDC systems [12–16]. Major research effort has been focused on the stable operation of the MIDC systems by introducing novel indexes, such as the multi-infeed interaction factor (MIIF), multi-infeed effective short-circuit ratio (MIESCR) [12], apparent increase in short-circuit ratio (AISCR) [13], and improved effective short-circuit ratio (IESCR) [14]. In parallel with such research, some studies have concentrated on the economic operation of MIDC systems, which is represented by optimal power flow (OPF) [17–20]. In two such studies [17,18], steady-state VSC–HVDC modeling methods and OPF formulations based on the Newton–Raphson method are proposed. Although only a single HVDC system is considered in these studies [17,18], the principle can be easily expanded to MIDC systems. In one study [19], a loss minimization method with an AC/DC hybrid grid incorporating only VSC HVDCs is proposed, based on an interior-point method. In another study [20], a similar method adopting both LCC and VSC HVDC is proposed. However, the authors of the above studies only consider the steady-state values of the HVDC systems. Thus, if the load profiles were different from the forecasted value, the previous methods were unable to find the optimal set-point, because the forecasting error was not considered.

A common method used to satisfy the power balance between generation and load caused by the forecasting error is frequency–power droop control [21]. Similar to conventional generators, HVDC systems also adopt frequency–power droop controllers [22–24]. For economic operation of power systems during frequency control, the calculation of droop coefficients is a significant issue, and the coefficients are used to exploit multiple HVDC systems efficiently. It is practically important in the case of an islanded network, because the islanded system has a small load level and high renewable energy penetration, which cause high uncertainties in the power system. Due to those characteristics, off-nominal frequency situations occur more frequently, and the frequency deviation is more severe than the conventional large power network. In one study [25], an optimum calculation method of voltage–power droop coefficients for multi-terminal HVDC (MTDC) systems is proposed. However, the method could only be applied to the specific topology of the DC grid investigated in the paper. Another study proposes an optimization-based droop coefficient calculation method to maximize converter efficiencies for a low-voltage system [26]. A method to minimize DC transmission line loss using voltage–current droop coefficients has also been proposed [27]. The above research suggests that the droop coefficients of converters can be calculated by optimization of the problem. However, to our knowledge, optimization problems have rarely been suggested to calculate frequency–power droop coefficients of MIDC systems.

This paper suggests a method to determine the frequency–power droop coefficients of multiple HVDCs in MIDC systems. First, we propose an optimization problem for MIDC systems, to determine the operating points of HVDCs to minimize system loss. Using the results of the optimization problem, we propose a droop coefficient design method based on Monte Carlo sampling. The proposed method is verified by MATLAB simulation utilizing IEEE test systems and the actual networks of Jeju Island in Korea. Using our method, a system operator can define the operating points and droop coefficients of an MIDC system more efficiently.

2. Description of the Proposed Method

2.1. Optimazation Problem to Determine Operating Points

In this section, we formulate an optimization problem to determine the operating points of multiple HVDCs, including LCC and VSC HVDCs. This study is different from previous studies covering various type of HVDCs [22–25,27], because HVDCs can be regarded as active and reactive power sources in MIDC systems. Also, only the outputs of multiple HVDCs are considered in the optimization problem. In other words, generator outputs are considered as a constant in the problem.

2.1.1. Model Description

To formulate the constraints used in the optimization problem, power balance equations and converter models are required. In the case of the VSC, which can synthesize sine-wave AC voltage regardless of the residual network, the active and reactive power output—P_{VSC} and Q_{VSC}, respectively—are regulated independently [28]. Therefore, a VSC can be considered similar to a *PQ* load for the optimization, and the only consideration is the rated output power, as follows:

$$\sqrt{P_{VSC}^2(n) + Q_{VSC}^2(n)} < S_{VSC.rate}(n), \quad (1)$$

where $S_{VSC.rate}(n)$ is the rated apparent power of the VSC HVDC connected to the n-th bus. Note that conversion loss of a VSC HVDC is ignored, because the input variables of controller are active and reactive power output at the inverter side. However, for future work, the conversion loss of VSC should be considered for solving overall optimization problems, including a DC system. The reactive power output of the LCC, Q_{LCC}, is different from the VSC, and cannot be regulated independent of the active power, P_{LCC} [29]. As conversion loss of LCC is less than 1% in general, the loss can be ignored [30]. Thus, Q_{LCC} can be presented as:

$$Q_{LCC}(n) = P_{LCC}(n) \tan \Phi(n) \quad (2)$$

where $\Phi(n)$ is the power factor angle of LCC HVDC connected to the n-th bus. The power factor angle can be expressed as follows [31]:

$$\tan \Phi(n) = \frac{2\mu(n) + \sin(2\alpha(n)) - \sin(2\alpha(n) + 2\mu(n))}{\cos(2\alpha(n)) - \cos(2\alpha(n) + 2\mu(n))} \quad (3)$$

where

$$\alpha(n) = \cos^{-1}\left[\frac{\pi}{3\sqrt{2}B(n)T(n)V(n)}\left(V_{DC.rate}(n) + \frac{3X_C(n)P_{LCC}(n)}{\pi V_{DC.rate}(n)}\right)\right] \quad (4)$$

$$\mu(n) = \cos^{-1}\left[\cos(\alpha(n)) - \frac{\sqrt{2}X_C(n)P_{LCC}(n)}{B(n)T(n)V(n)V_{DC.rate}}\right] - \alpha \quad (5)$$

where $V(n)$ represents voltage magnitude at n-th bus, and $V_{DC.rate}(n)$ is the rated DC voltage of the LCC HVDC connected to the n-th bus. The transformer turn ratio and the number of six-pulse bridges at the n-th bus are represented by $T(n)$ and $B(n)$, respectively. The reactance of the converter transformer at the n-th bus is described by $X_C(n)$. An inequality constraint of an LCC HVDC can be represented as

$$P_{LCC}(n) < P_{LCC.rate}(n) \quad (6)$$

where $P_{LCC.rate}(n)$ is the rated active power of the LCC HVDC connected to n-th bus. Note that the index of the bus number (n) is utilized to express HVDC system connected to the n-th bus, which means multiple HVDCs can be considered in the proposed models, because each converter is connected to a different bus.

For every bus, the injected active (P) and reactive (Q) power, including generator output (P_{GEN} and Q_{GEN}), VSC output (P_{VSC} and Q_{VSC}), LCC output (P_{LCC} and Q_{LCC}), and connected load (P_{LOAD} and Q_{LOAD}) satisfy the following constraints [32]:

$$P_{GEN}(n) + P_{VSC}(n) + P_{LCC}(n) - P_{LOAD}(n) - \sum_{k=1}^{N} V(n)V(k)\{G(n,k)\cos(\theta(n) - \theta(k)) + B(n,k)\sin(\theta(n) - \theta(k))\} = 0 \quad (7)$$

$$Q_{GEN}(n) + Q_{VSC}(n) + Q_{LCC}(n) - Q_{LOAD}(n) - \sum_{k=1}^{N} V(n)V(k)\{G(n,k)\sin(\theta(n) - \theta(k)) - B(n,k)\cos(\theta(n) - \theta(k))\} = 0 \quad (8)$$

where the phase angle of n-th bus is represented by $\theta(n)$. An injected active and reactive power of the n-th bus can be represented by $P(n)$ and $Q(n)$, respectively. The real and imaginary parts of the admittance matrix between the n-th bus and k-th bus are represented as $G(n,k)$ and $B(n,k)$, respectively. If the generator, VSC, LCC, or load is not connected to bus n, the value of each symbol is zero. Note that P_{GEN} and Q_{GEN} are pre-determined by the system operator, and P_{LOAD} and Q_{LOAD}

are previously forecasted values, which are considered as constant in the optimization problem. The security constraint of the AC voltage magnitude for per unit (p.u.) system can be represented as:

$$0.95 \text{ p.u.} < V(n) < 1.05 \text{ p.u.} \tag{9}$$

2.1.2. Optimization Formulation and Solving Method

The vectors of the phase angle and voltage magnitude are represented as $\boldsymbol{\theta}$ and $\mathbf{V}$, respectively. The output references of VSC and LCC HVDCs can be represented by vector form, as $\mathbf{P_{VSC}}$, $\mathbf{Q_{VSC}}$, and $\mathbf{P_{LCC}}$. Note that the reactive power of LCC HVDC, $\mathbf{Q_{LCC}}$, can be represented in terms of $\mathbf{P_{LCC}}$ and $\mathbf{V}$ using Equations (2)–(5), so it is not necessary to include $\mathbf{Q_{LCC}}$ with the unknown vector. Therefore, the unknown vector, $\mathbf{X}$, can be defined as:

$$\mathbf{X} = \begin{bmatrix} \mathbf{P}_{\mathbf{VSC}}^T & \mathbf{Q}_{\mathbf{VSC}}^T & \mathbf{P}_{\mathbf{LCC}}^T & \boldsymbol{\theta}^T & \mathbf{V}^T \end{bmatrix}^T. \tag{10}$$

The purpose of the proposed problem is to minimize the system loss. Since conversion loss is not included in MIDC systems, only the line loss of AC systems is considered in the objective function. The objective function of the optimization is represented as:

$$\text{minimize } f = \sum_{j=n}^{N} \{P_{GEN}(n) + P_{VSC}(n) + P_{LCC}(n) - P_{LOAD}(n)\} \tag{11}$$

where N is the number of buses. Note that the difference between the active power injected to the network, P_{GEN}, P_{VSC}, P_{LCC}, and P_{LOAD} is the same as the system loss. Only LCC and VSC HVDCs are considered as frequency-supporting resources in this paper, with the assumption that conventional generators do not participate in frequency support. However, conventional generators can be considered in the optimization problem with a little modification, because the generators can be represented as active and reactive power sources similar to the VSC HVDC system. Additionally, the security constraint of the conventional generators can be represented by simple inequality of active and reactive power [33].

To solve the optimization problem, an interior-point method is used [34,35]. For the proposed optimization problem, an interior-point method cannot guarantee global optimality. However, as the solution found by the method is better than a trivial solution, the interior-point method can be successfully applied to the proposed optimization problem.

2.2. Frequency–Power Droop Coefficient Determination Method

In the optimization problem described in the previous section, the forecasted load profile is used to calculate the optimal operating points. However, load characteristics are different from the forecasted values, because there are uncertainties in the load forecasting procedure. Furthermore, uncertainty increases with the integration of bulk renewable energies [36]. In this situation, the outputs of multiple HVDCs form new operating points following droop characteristics, the coefficient of which is generally proportional to the capacities of the converters [37]. The conventional method for determining droop coefficients is simple, but the coefficients cannot create a more efficient solution. Therefore, a droop coefficient design method based on an optimization problem is required to operate MIDC systems more economically. We propose a statistical method to obtain optimal droop coefficients in this section.

2.2.1. Stochastic Optimization Based on the Monte Carlo Sampling Method

To determine the droop coefficient in the proposed method, the Monte Carlo sampling method was utilized [38]. In the Monte Carlo sampling method, a number of load profiles are generated and are utilized for solving the optimization problem. All cases are generated based on probabilistic function, such as normal distribution, to represent possible realizations in the presence of forecast errors. For each load profile, the optimization problem of the previous section is solved, and operating points of HVDCs are calculated. The sampling number is represented by i, and the output of LCC and VSC HVDCs connected to the n-th bus at the i-th sample is represented as $P_{VSC}(n,i)$ and $P_{LCC}(n,i)$, respectively.

Furthermore, frequency deviation for each load profile is also calculated in this stage. To derive frequency deviation, the relationship between load profile and grid frequency should be established. Therefore, on the assumption that only HVDCs participate in frequency regulation, we use the swing equations of the power network, including frequency-supporting HVDCs and a load with damping effects, as follows [21]:

$$\sum_{j=n}^{N}\{P_{GEN}(n,i)+P_{VSC}(n,i)+P_{LCC}(n,i)-P_{LOAD}(n,i)\}=M_{eq}\frac{d\Delta\omega(i)}{dt}+D\Delta\omega(i) \tag{12}$$

where M_{eq} is the moment of inertia and D is the load damping coefficient. The deviation of grid frequency is represented by $\Delta\omega$. As the steady-state value (i.e., $\Delta\omega = 0$) of the left side of Equation (12) is equal to zero, and the active power of the HVDCs can be represented by the corresponding droop coefficients, the frequency deviation can be described as

$$\Delta\omega(i)=\frac{-\sum\limits_{n=1}^{N}\Delta P_{LOAD}(n,i)}{\sum\limits_{n=1}^{N}\frac{1}{R_{VSC}(n)}+\sum\limits_{n=1}^{N}\frac{1}{R_{LCC}(n)}+D}, \tag{13}$$

where $\Delta P_{LOAD}(n,i)$ is the change in the load profile of the n-th bus at the i-th sample, which is defined by the Monte Carlo sampling method. $R_{VSC}(n)$ and $R_{LCC}(n)$ represent the droop coefficient of the VSC and LCC HVDC connected to the n-th bus, respectively. The frequency deviation cannot be derived directly from Equation (13), because the droop coefficients of the HVDCs, which are not determined yet, are utilized to derive the frequency deviation. Therefore, we assume that the sum of coefficients is constant, as follows:

$$\sum_{n=1}^{N}\frac{1}{R_{VSC}(n)}+\sum_{n=1}^{N}\frac{1}{R_{LCC}(n)}=\frac{1}{R_{eq}} \tag{14}$$

where R_{eq} is an equivalent droop coefficient of the total system. Note that the active power of HVDCs during grid frequency only depends on the ratio of the droop coefficient, so power sharing between multiple HVDCs can be regulated properly while satisfying the constraint (14).

2.2.2. Graphical Analysis to Determine Droop Coefficients

In this section, we derive the droop coefficients of LCC and VSC HVDCs by graphical analysis. Figure 1 shows a concept of the proposed method of droop coefficient determination. Red points are operating points of LCC and VSC HVDCs determined by the optimization problem. Grey dots represent optimization results of the Monte Carlo samples. The points are categorized by active power reference and grid frequency, so that the frequency–power droop can be represented by the slope of the blue line, to minimize the root mean square error between the points above the blue line and the grey dots. The slopes of the blue lines are presented as $-R_{VSC,0}$ for VSC HVDC and $-R_{LCC,0}$ for the LCC HVDC. Note that the determination method is called the "graphic method", in that curve-fitting is exploited for coefficient determination.

We can derive droop coefficients of LCC and VSC HVDCs from Figure 1, but the coefficients may not satisfy Equation (14), because the power loss of the system is not considered in (12). We can determine final values of the coefficients by the additional correction method. As active power sharing ratio of HVDCs depend on the ration of droop coefficients, final droop coefficients are corrected by multiplying the same constant to $R_{LCC,0}$ and $R_{VSC,0}$. Thus final coefficients are determined as

$$R_{VSC}(k) = \left(\sum_{n=1}^{N} \frac{R_{eq}}{R_{VSC,0}(n)} + \sum_{n=1}^{N} \frac{R_{eq}}{R_{LCC,0}(n)} \right) R_{VSC,0}(k) \;\&\; R_{LCC}(k) = \left(\sum_{n=1}^{N} \frac{R_{eq}}{R_{VSC,0}(n)} + \sum_{n=1}^{N} \frac{R_{eq}}{R_{LCC,0}(n)} \right) R_{LCC,0}(k), \quad (15)$$

where $R_{VSC,0}(n)$ and $R_{LCC,0}(n)$ are the coefficients of the HVDCs connected to n-th bus, derived from the graphical analysis.

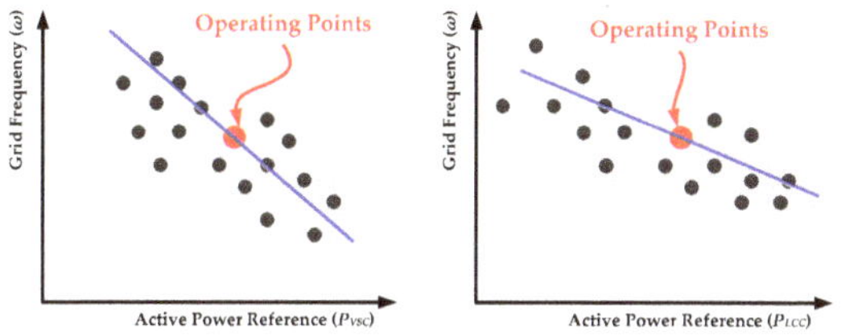

Figure 1. Concept of the method to determine frequency–power droop coefficients of voltage sourced converter (VSC) and line commutated converter (LCC) high-voltage direct currents (HVDCs).

2.3. Overall Procedure

Figure 2 shows the overall procedure of the proposed method. First, the operating points of multiple HVDCs are derived from the forecasted load and scheduled output of generators using Equation (11). The Monte Carlo sampling method extracts irregular load profiles, and the active power references and frequency deviation for all samples are calculated by the samples. Note that the flow of multiple scenarios is represented by a dotted line. Finally, using the graphical approach described in Figure 1, the frequency–power droop coefficients of VSC HVDCs and LCC HVDCs are determined.

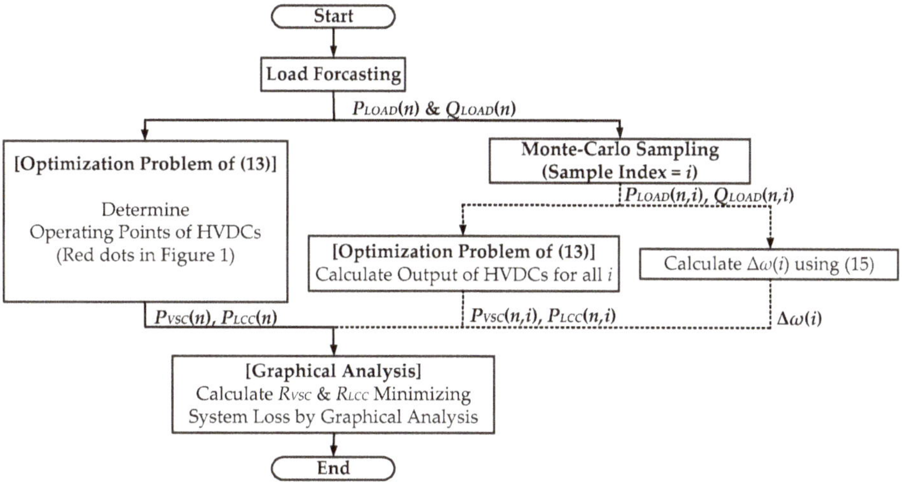

Figure 2. Overall procedure of the proposed method to calculate droop coefficients.

3. Simulation Results

Case studies were performed using two test systems. First, we verified the proposed method using the IEEE 14-bus test system, which is widely utilized to evaluate algorithms considering transmission networks [39]. Then, we provided the simulation results considering an actual power system, i.e., that of Jeju Island in Korea.

3.1. Simulation Results for the IEEE 14-Bus Test System

Figure 3 represents the modified IEEE 14-bus test system, including both LCC and VSC HVDC systems. The base of the complex power is 100 MW. Synchronous condensers are eliminated from the original test system, and only a single generator at Bus #1 is considered for simplification. Because Bus #1 is the reference bus, and the voltage-controlling generator is installed at the bus, the voltage magnitude and angle are 1 p.u. and 0 rad, respectively. The active power of the generator is fixed at 50 MW (0.5 p.u.) because it is not considered as a variable in the optimization problem. A 500 kV/200 MW LCC HVDC and a 400 kV/200 MW VSC HVDC are included at Bus #2 and #3, respectively. The parameters of the LCC HVDC are modified from the well-known CIGRE BENCHMARK model [40]. In the test system, the variables to be determined are P_{LCC}, P_{VSC}, and Q_{VSC}. Note that the harmonic filters and shunt capacitors of the LCC HVDC system are considered at the admittance matrix.

We compared the steady-state results of two conventional methods and the proposed method. In the first method, VSC HVDC operates in unity power factor (UPF) mode, so that the reactive power output is zero. In the second method (VC mode), the VSC HVDC controls AC voltage to be as high as possible. In both methods, the active power references of LCC and VSC HVDCs are identical. The forecasted load profile in the test system is illustrated in Figure 4, which is applied to both case studies.

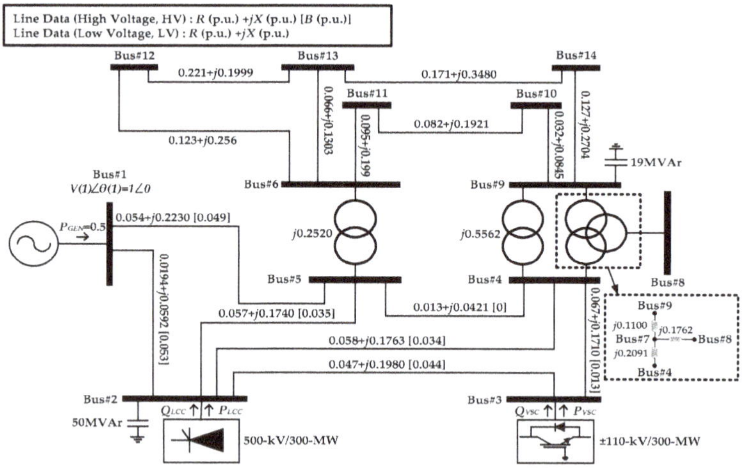

Figure 3. Configuration of Institute of Electrical and Electronics Engineers (IEEE) 14-bus test system.

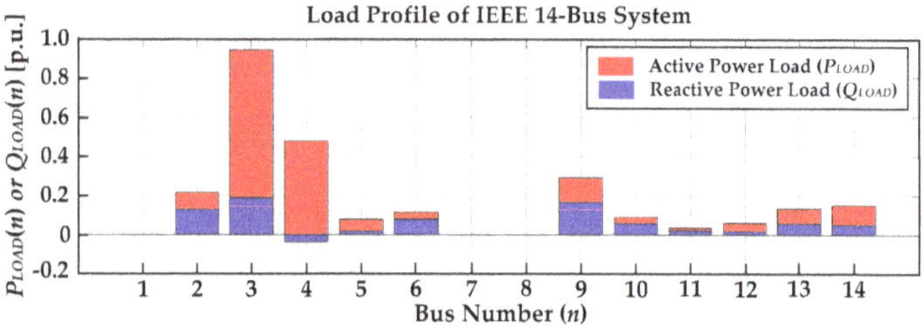

Figure 4. Forecasted load profile for the IEEE 14-bus test system.

Figure 5 shows voltage profiles with the conventional and proposed method. Voltage profiles are investigated in the case studies, in that security constraints of AC voltage should be satisfied by utilizing the optimization problem. In UPF mode, the AC voltage profiles of most buses are lower than the lower bound (V_{lb}), as shown by the blue areas of Figure 5. On the other hand, in VC mode, AC voltage is maintained within the operational boundaries, because VSC HVDC compensates reactive power into the network as much as possible. The voltage is also maintained within its boundaries using the proposed method, because constraints on AC voltage are reflected in the optimization problem. However, the amount of reactive power with the proposed method is smaller than that of the VC mode, as shown in Figure 5.

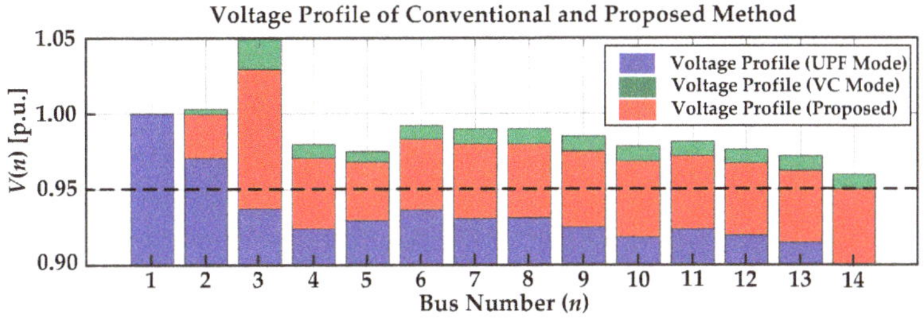

Figure 5. Voltage profile of the conventional and proposed methods in the IEEE 14-bus test system.

Table 1 shows the comparisons of active power outputs of LCC and VSC HVDCs, and the resulting power loss in three cases. As the outputs of LCC and VSC HVDCs are intended to be identical in the conventional methods, both HVDCs provide the same active power into the power network. On the other hand, VSC HVDC provides more active power than LCC HVDC in the case of the proposed method to minimize system loss. As a result, the active power loss of the test system with the proposed method is reduced 13.89% and 11.79% from that of the UPF mode and VC mode, respectively. Even the voltage level is higher when VSC HVDC operates on VC mode than the proposed case; power loss is smaller when using the proposed method, because the current flow increases for reactive power compensation in VC mode. As shown in Figure 5 and Table 1, the simulation results suggest that the proposed method is effective with steady-state characteristics for the IEEE 14-bus test system.

Table 1. Comparison of power in three cases.

	UPF Mode	VC Mode	Proposed
Output of LCC HVDC (P_{LCC})	112.02 MW	111.96 MW	84.08 MW
Output of VSC HVDC (P_{VSC})	112.02 MW	111.96 MW	139.26 MW
Power loss	5.04 MW	4.92 MW	4.34 MW

UPF: unity power factor, **VC**: voltage control.

To verify the droop coefficient determination method, we generate 10,000 load profiles reflecting forecasting error. We assume that the active and reactive load profile follows the normal distribution function, in which the average is the forecasted value and the standard deviation is 3% of the forecasted value. According to a previous study [21], we define R_{eq} and D as 0.1 and 2, respectively. Figure 6 illustrates the graphical analysis of the frequency–power droop coefficient design methodology. As the optimization results are scattered as grey dots in Figure 6, we can define the droop coefficients of LCC and VSC HVDCs—R_{LCC} and R_{VSC}, respectively—like their corresponding lines.

Using Equation (15), we can derive the final droop coefficient values as R_{LCC} = 0.227 and R_{VSC} = 0.179. The calculated droop coefficients are verified by comparison with the conventional case. The conventional case is defined as the situation where the operating points are determined by the optimization, and the droop coefficients are determined proportional to their capacity [41]. To verify the proposed determination method, we utilized 10 randomly sampled load profiles. Table 2 shows the comparison of average outputs and power losses. As described in Table 2, the deviation of power loss increased using the proposed method to reduce total system loss. From the simulation results in Tables 1 and 2, we suggest that the proposed optimization problem and the droop coefficient determination method enhance the efficiency of the test system.

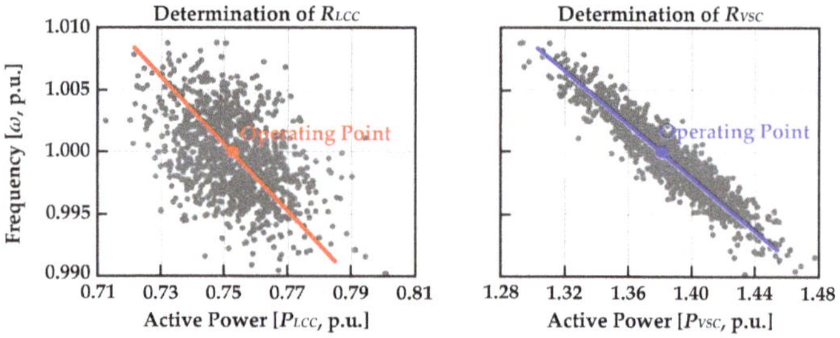

Figure 6. Graphical coefficient determination method applied to the IEEE 14-bus test system.

Table 2. Comparison of power loss between randomly sampled load profiles.

	Conventional	Proposed
Deviation of LCC HVDC's output (ΔP_{LCC})	−0.32 MW	−0.28 MW
Deviation of VSC HVDC's output (ΔP_{VSC})	−0.32 MW	−0.35 MW
Deviation of power loss	−36.41 kW	−36.74 kW

3.2. Simulation Results for the Jeju Island System

Figure 7 shows the configuration of the Jeju Island power network [42]. There are four generators and three HVDC systems, i.e., two LCC HVDCs and one VSC HVDC. The first LCC HVDC is a 180 kV/300 MW system constructed in 1998, and the second LCC HVDC is a ±250 kV/400 MW system constructed in 2013. The VSC HVDC considered in the simulation is a planned installation with a capacity of ±150 kV/200 MW. The base of complex power is 100 MW, and the base voltage is 154 kV. The active power of generators, line impedance, and data of the shunt compensator in the test system are represented in Appendix A. The actual load data from Jeju Island at peak time is utilized, as illustrated in Figure 8. The simulation scenarios are similar to those with the IEEE 14-bus test system. Only one generator at Jeju 3C/S controls the voltage magnitude at 1.04 p.u., and the other generators only regulate active power, which means that the reactive power of these generators is zero. The output of the conventional method is determined proportional to capacity.

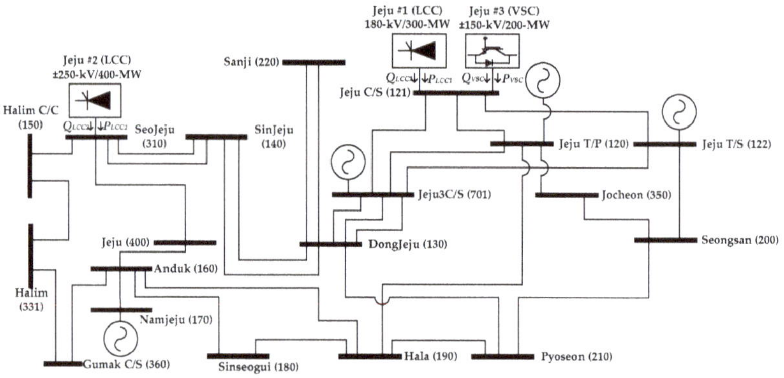

Figure 7. Configuration of the Jeju Island network with multiple HVDCs.

Table 3 shows the results of steady-state analysis using the conventional and proposed methods. As the outputs of multiple HVDCs are determined proportional to their capacity in the conventional method, the active power ratio of Jeju #1, #2, and #3 is 3:4:2. On the other hand, the results of the proposed method determine the operating points of the HVDCs using an optimization problem. Thus, power loss in the system was reduced 13.82% from that of the conventional method. The results suggest that operation with the proposed optimization problem would reduce power loss. In the second case, Jeju #1 and Jeju #3 are connected to same bus. Among two HVDC systems, the ratio of active power is not significant, because they are connected to same bus. However, the references to them are determined because LCC HVDC absorbs reactive power different from the VSC HVDC. Therefore, the proposed optimization problem considers not only active power but also optimal reactive power flow.

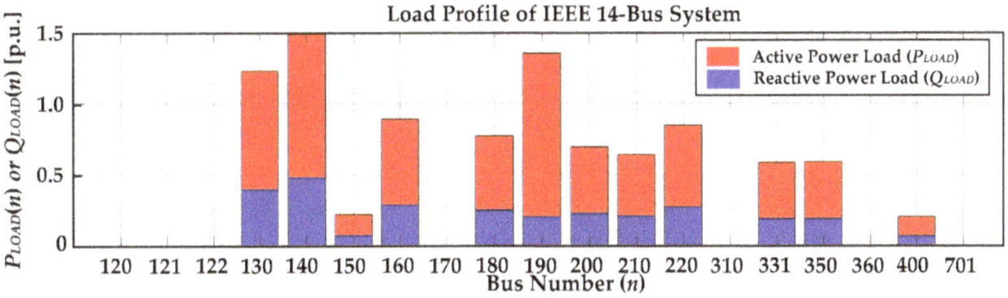

Figure 8. Load profile for Jeju Island.

Table 3. Comparison of power loss between the conventional and proposed methods.

	Conventional	Proposed
Output of Jeju #1 (P_{LCC1})	132.82 MW	50.17 MW
Output of Jeju #2 (P_{LCC2})	177.09 MW	305.73 MW
Output of Jeju #3 (P_{VSC})	88.55 MW	41.80 MW
Power loss	6.26 MW	5.50 MW

Figure 9 presents a graphical analysis of droop coefficient determination. To verify the droop coefficient determination method, 1000 randomly extracted scenarios were utilized. Equivalent droop coefficients R_{eq} and D were assumed to be identical to the previous case study. As a result, we could derive the droop coefficients, R_{LCC1}, R_{LCC2}, and R_{VSC}, as 0.298, 0.233, and 0.426, respectively.

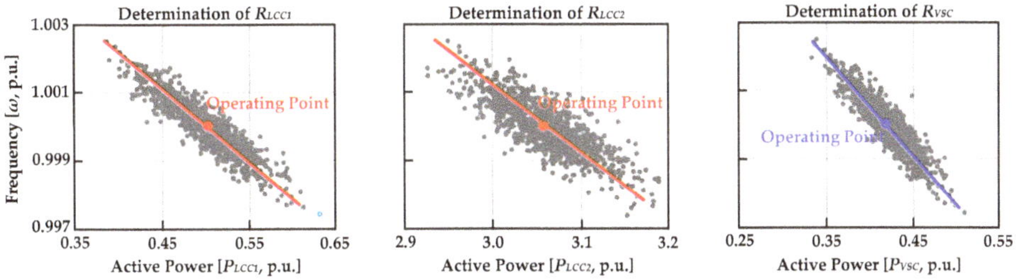

Figure 9. Graphical analysis of the coefficient determination method applied to the Jeju Island system.

The droop coefficients derived from Figure 9 are verified by comparison with the conventional case, in which the steady-state system is determined by an optimization problem. The only difference between the conventional and proposed methods is that the droop coefficients are proportional to capacity in the conventional method. Table 4 shows the average loss deviation from the power loss at steady-state for 1000 random cases. Power loss deviation is reduced 0.48% using the proposed droop coefficients. The difference is very small because the droop coefficient in the conventional method is very similar to that of the proposed method; however, the power loss is successfully reduced using the proposed method.

Table 4. Comparisons of average power loss deviation for irregular load profiles.

	Conventional	Proposed
Deviation of power loss	1.683 kW	1.675 kW

4. Conclusions

We propose the use of an optimization problem for MIDC systems. The optimization problem considers both the LCC and VSC HVDCs, and we derived their models to analyze the efficiency of the AC network. The purpose of the problem is loss minimization. Furthermore, we utilized the optimization problem to determine the frequency–power droop coefficients of multiple HVDCs, considering characteristics of both the LCC and VSC HVDCs. The determination method calculates the coefficients by minimizing the root mean square error between the numerous optimization results from the Monte Carlo sampling method, asw well as the linear-approximated values. As several small optimization problems are solved in the proposed method rather than one big problem, the proposed method reduces computation burdens compared with conventional stochastic optimization problems. In other words, the computation time increases arithmetically with the number of samples for our method. Two test systems, an IEEE 14-bus test system and an actual Jeju grid, were used to verify the proposed method. From the simulation results, we suggest that the proposed method efficiently reduces loss in the power system. For future work, consideration of an overall system, including HVDC networks and an AC system for the sending side, is required. By investigating the overall system using the proposed method, we can find globally optimized operating points of multiple HVDC systems, in order to enhance power system efficiency.

Author Contributions: The main optimization algorithm and the droop coefficient determination method were proposed by G.L. The test system data were collected and analyzed by P.H. The entire article and the simulation results were reviewed by S.M.

Funding: This research received no external funding.

Acknowledgments: This work was supported by the Human Resources Development program of the Korea Institute of Energy Technology Evaluation and Planning (KETEP) grant, funded by Korean government Ministry of Trade, Industry, and Energy (No. 20174030201540).

Conflicts of Interest: The authors declare no conflict of interest.

Appendix A

Table A1. Bus data of the Jeju Island network.

Number	Name	P_{GEN} [MW]	Q_{SHUNT} [MVAr]	P_{LOAD} [MW]	Q_{LOAD} [MVAr]
120	Jeju T/P	200.1	−24.4	0	0
121	Jeju C/S	0	100	0	0
122	Jeju T/S	75.4	−14.2	0	0
130	Dongjeju	0	0	123.1	39.1
140	Sinjeju	0	0	149.3	47.5
150	Halim C/C	0	0	21.8	6.9
160	Anduk	0	10	89	28.3
170	Namjeju	88.3	30.8	0	0
180	Sinseogui	0	0	77	24.5
190	Hala	0	0	135.2	20
200	Seongsan	0	0	69.4	22.1
210	Pyoseon	0	0	64.1	20.4
220	Sanji	0	0	84.6	26.9
310	Seojeju	0	168	0	0
331	Halim	0	0	58.5	18.6
350	Jocheon	0	0	59	18.8
360	Gumak C/S	0	34	0	0
400	Jeju	0	0	20	6.4
701	Jeju3C/S	195	0	0	0

Table A2. Branch data of the Jeju Island network.

To	From	Id	R [p.u.]	X [p.u.]	B [p.u.]
Jeju T/P	Jeju C/S	1	0	0.00001	0
Jeju T/P	Hala	1	0.012046	0.058234	0.089314
Jeju T/P	Jocheon	1	0.008051	0.038281	0.028533
Jeju T/P	Jeju3C/S	1	0.000583	0.006068	0.171872
Jeju C/S	Jeju T/S	1	0	0.00001	0
Jeju C/S	Dongjeju	1	0.005688	0.026864	0.010699
Jeju T/S	Seongsan	1	0.012916	0.060538	0.030189
Jeju T/S	Jeju3C/S	1	0.000586	0.006004	0.171872
Dongjeju	Sinjeju	1	0.007066	0.033294	0.013268
Dongjeju	Sinjeju	2	0.007066	0.033294	0.013268
Dongjeju	Pyoseon	1	0.014445	0.071026	0.142377
Dongjeju	Sanji	1	0.000404	0.004380	0.087103
Dongjeju	Sanji	2	0.000404	0.004380	0.087103
Dongjeju	Jeju3C/S	1	0.000583	0.006068	0.171872
Dongjeju	Jeju3C/S	2	0.000583	0.006068	0.171872
Dongjeju	Jeju3C/S	3	0.000583	0.006068	0.171872
Sinjeju	Seojeju	1	0.001280	0.005670	0.002459
Sinjeju	Seojeju	2	0.001280	0.005670	0.002459
Halim C/C	Seojeju	1	0.012159	0.053845	0.023354
Halim C/C	Halim	1	0.000110	0.000100	0
Anduk	Namjeju	1	0.000759	0.006820	0.121576
Anduk	Namjeju	2	0.000763	0.006810	0.121638
Anduk	Sinseogui	1	0.009642	0.044350	0.019041
Anduk	Hala	1	0.014618	0.066670	0.028975
Anduk	Gumak C/S	1	0.006367	0.029764	0.012026
Anduk	Jeju	1	0.007632	0.035500	0.014904
Sinseogui	Hala	1	0.008861	0.0740814	0.017478
Hala	Pyoseon	1	0.009614	0.046274	0.067795
Seongsan	Pyoseon	1	0.001086	0.006344	0.190712
Seongsan	Jocheon	1	0.009854	0.046712	0.031940
Seojeju	Jeju	1	0.002120	0.009875	0.004140
Halim	Gumak C/S	1	0.002037	0.009525	0.003848

References

1. Figueroa-Acevedo, A.L.; Czahor, M.S.; Jahn, D.E. A comparison of the technological, economic, public policy, and environmental factors of HVDC and HVAC interregional transmission. *AIMS Energy* **2015**, *3*, 144–161. [CrossRef]
2. Li, B.; Liu, T.; Xu, W.; Li, Q.; Zhang, Y.; Li, Y.; Li, X.Y. Research on technical requirements of line-commutated converter-based high-voltage direct current participating in receiving end AC system's black start. *IET Gener. Transm. Distrib.* **2016**, *10*, 2071–2078. [CrossRef]
3. Wang, L.; Thi, M.S. Stability enhancement of a PMSG-based offshore wind farm fed to a multi-machine system through LCC-HVDC link. *IEEE Trans. Power Syst.* **2013**, *28*, 3327–3334. [CrossRef]
4. Bidadfar, A.; Nee, H.; Zhang, L.; Harnefors, L.; Namayantavana, S.; Abedi, M.; Karrari, M.; Gharehpetian, G.B. Power system stability analysis using feedback control system modeling including HVDC transmission links. *IEEE Trans. Power Syst.* **2016**, *31*, 116–124. [CrossRef]
5. Azad, S.P.; Taylor, J.A.; Iravani, R. Decentralized supplementary control of multiple LCC-HVDC links. *IEEE Trans. Power Syst.* **2016**, *31*, 572–580. [CrossRef]
6. Azad, S.P.; Iravani, R.; Tate, J.E. Stability enhancement of a DC-segmented AC power system. *IEEE Trans. Power Deliv.* **2015**, *30*, 737–745. [CrossRef]
7. Kwon, D.; Kim, Y.; Moon, S. Modeling and Analysis of an LCC HVDC system using DC voltage control to improve transient response and short-term power transfer capability. *IEEE Trans. Power Deliv.* **2018**, *33*, 1922–1933. [CrossRef]
8. Zhang, M.; Yuan, X.; Hu, J. Inertia and primary frequency provisions of PLL-synchronized VSC HVDC when attached to islanded AC system. *IEEE Trans. Power Syst.* **2018**, *33*, 4179–4188. [CrossRef]

9. Jang, G.; Oh, S.; Han, B.M.; Kim, C.K. Novel reactive power compensation scheme for the Jeju-Haenam HVDC system. *IEE Proc.-Gener. Transm. Ditrib.* **2005**, *152*, 514–520. [CrossRef]
10. Market, P.E.; Skliutas, J.P.; Sung, P.Y.; Kim, K.S.; Kim, H.M.; Sailer, L.H.; Young, R.R. New synchronous condensers for Jeju island. In Proceedings of the 2012 IEEE Power & Energy Society General Meeting, San Diego, CA, USA, 22–26 July 2012.
11. Yoon, M.; Yoon, Y.; Jang, G. A study on maximum wind power penetration limit in island power system considering high-voltage direct current interconnections. *Energies* **2015**, *8*, 14244–14259. [CrossRef]
12. Davies, B.; Williamson, A.; Gole, A.M.; Ek, B.; Long, B.; Burton, B.; Kell, D.; Brandt, D.; Lee, D.; Rahimi, E.; et al. Systems with multipld DC infeed. In *CIGRE Working Group B4.41*; CIGRE: Paris, France, 2008.
13. Gui, C.; Zhang, Y.; Gole, A.M.; Zhao, C. Analysis of dual-infeed HVDC with LCC-HVDC and VSC-HVDC. *IEEE Trans. Power Deliv.* **2012**, *27*, 1529–1597. [CrossRef]
14. Ni, X.; Gole, A.M.; Zhoa, C.; Guo, C. An improved measure of AC system strength for performance analysis of multi-infeed HVdc systems including VSC and LCC converters. *IEEE Trans. Power Deliv.* **2018**, *33*, 169–178. [CrossRef]
15. Rahimi, E.; Gole, A.M.; Davies, J.B.; Fernando, I.T.; Kent, K.L. Commutation failure analysis in multi-infeed HVDC systems. *IEEE Trans. Power Deliv.* **2011**, *26*, 378–384. [CrossRef]
16. Hwang, S.; Lee, J.; Jang, G. HVDC-system-interaction assessment through line flow change distribution factor and transient stability analysis at planning stage. *Energies* **2016**, *9*, 1068. [CrossRef]
17. Rabiee, A.; Soroudi, A.; Keane, A. Information gap decision theory based OPF with HVDC connected wind farms. *IEEE Trans. Power Syst.* **2015**, *30*, 3396–3406. [CrossRef]
18. Pizano-Martinez, A.; Fuerte-Esquivel, C.R.; Ambriz-Perez, H.; Acha, E. Modeling of VSC-based HVDC systems for a Newton-Raphson OPF algorithm. *IEEE Trans. Power Syst.* **2007**, *22*, 1794–1803. [CrossRef]
19. Gavriluta, C.; Candela, I.; Luna, A.; Gomez-Exposito, A.; Rodriguez, P. Hierarchical control of HV-MTDC systems with droop-based primary and OPF-based secondary. *IEEE Trans. Smart Grid* **2015**, *6*, 1502–1510. [CrossRef]
20. Han, M.; Xu, D.; Wan, L. Hierarchical optimal power flow control for loss minimization in hybrid multi-terminal HVDC transmission system. *CSEE J. Power Energy Syst.* **2016**, *2*, 40–46. [CrossRef]
21. Kundur, P.; Balu, N.J.; Lauby, M.G. *Power System Stability and Control*; McGraw-Hill: New York, NY, USA, 1994.
22. Liu, H.; Chen, Z. Contribution of VSC-HVDC to frequency regulation of power systems with offshore wind generation. *IEEE Trans. Energy Convers.* **2015**, *30*, 918–926. [CrossRef]
23. Adeuyi, O.D.; Cheah-Mane, M.; Liang, J.; Jenkins, N. Fast frequency response from offshore multiterminal VSC-HVDC schemes. *IEEE Trans. Power Deliv.* **2017**, *32*, 2442–2452. [CrossRef]
24. Kwon, D.; Kim, Y.; Moon, S.; Kim, C. Modeling of HVDC system to improve estimation of transient DC current and voltages for AC line-to-ground fault-an actual case study in Korea. *Energies* **2017**, *10*, 1543. [CrossRef]
25. Abdel-Khalik, A.S.; Massoud, A.M.; Elserougi, A.A.; Ahmed, S. Optimum power transmission-based droop control design for multi-terminal HVDC of offshore wind farm. *IEEE Trans. Power Syst.* **2013**, *28*, 3401–3409. [CrossRef]
26. Agundis-Tinajero, G.; Diaz, N.L.; Luna, A.C.; Segundo-Ramírez, J.; Visairo-Cruz, N.; Gerrero, J.M.; Vazquez, J.C. Extended-optimal-power-flow-based hierarchical control for islanded AC microgrids. *IEEE Trans. Power Electron.* **2018**, *34*, 840–848. [CrossRef]
27. Cao, J.; Du, W.; Wang, H.F.; Bu, S.Q. Minimization of transmission loss in meshed AC/DC grids with VSC-MTDC networks. *IEEE Trans. Power Syst.* **2013**, *28*, 3047–3055. [CrossRef]
28. Flourentzou, N.; Agelidis, V.G.; Demetriades, G.D. VSC-based HVDC power transmission systems: An overview. *IEEE Trans. Power Electron.* **2009**, *24*, 592–602. [CrossRef]
29. Li, Y.; Luo, L.; Rehtanz, C.; Rüberg, S.; Liu, F. Realization of reactive power compensation near the LCC-HVDC converter bridges by means of an inductive filtering method. *IEEE Trans. Power Electron.* **2009**, *24*, 592–602. [CrossRef]
30. He, X.; Geng, H.; Yang, G.; Zou, X. Coordinated Control for Large-Scale Wind Farms with LCC-HVDC Integration. *Energies* **2018**, *11*, 2207. [CrossRef]
31. DJesus, M.E.M.; Martin, D.S.; Arnaltes, S.; Castronuovo, E.D. Optimal operation of offshore wind farms with line-commutated HVDC link connection. *IEEE Trans. Energy Convers.* **2010**, *25*, 504–513. [CrossRef]

32. Bergen, A.R.; Vittal, V. *Power Systems Analysis*, 2nd ed.; Prentice-Hall: Upper Sandle River, NJ, USA, 1994.
33. Sahli, Z.; Hamouda, A.; Bekrar, A.; Trentesaux, D. Reactive Power Dispatch Optimization with Voltage Profile Improvement Using an Efficient Hybrid Algorithm†. *Energies* **2018**, *11*, 2134. [CrossRef]
34. Jeong, M.G.; Kim, Y.J.; Moon, S.I.; Hwang, P.I. Optimal voltage control using an equivalent model of a low-voltage network accommodating inverter-interfaced distributed generators. *Energies* **2017**, *10*, 1180. [CrossRef]
35. Huang, Y.; Yang, K.; Zhang, W.; Lee, K.Y. Hierarchical energy management for the multienergy carriers system with different interest bodies. *Energies* **2018**, *11*, 2834. [CrossRef]
36. Li, J.; Wang, B.; Ren, H.; Zhao, D.; Wang, F.; Shafie-khah, M.; Catalão, J.P.S. Two-tier reactive power and voltage control strategy based on ARMA renewable power forecasting models. *Energies* **2017**, *10*, 1518.
37. Yuan, C.; Xie, P.; Yang, D.; Xiao, X. Transient stability analysis of islanded AC microgrids with a significant share of virtual synchronous generators. *Energies* **2018**, *11*, 44. [CrossRef]
38. Sauhats, A.; Zemite, L.; Petrichenko, L.; Moshkin, I.; Jasevics, A. Estimating the economic impacts of net metering schemes for residential PV systems with profiling of power demand, generation, and market prices. *Energies* **2018**, *11*, 3222. [CrossRef]
39. *Modeling and Simulation of IEEE 14 Bus System with FACTS Controllers*; Electrical and Computer Engineering Department, University of Waterloo: Waterloo, ON, Canada, 2003. Available online: https://www.researchgate.net/profile/Mohamed_Mourad_Lafifi/post/Datasheet_for_5_machine_14_bus_ieee_system2/attachment/59d637fe79197b8077995408/AS%3A395594351824896%401471328451959/download/MODELING+AND+SIMULATION+OF+IEEE+14+BUS+SYSTEM+WITH+FACTS+CONTROLLERS+IEEEBenchmarkTFreport.pdf (accessed on 11 February 2019).
40. Faruque, M.O.; Zhang, Y.; Dinavahi, V. Detailed modeling of CIGRE HVDC benchmark system using PSCAD/EMTDC and PSB/SIMULINK. *IEEE Trans. Power Deliv.* **2005**, *21*, 378–387. [CrossRef]
41. Mohamed, Y.A.-R.I.; El-Saadany, E.F. Adaptive decentralized droop controller to preserve power sharing stability of paralleled inverters in distributed generation microgrids. *IEEE Trans. Power Electron.* **2008**, *23*, 2806–2816. [CrossRef]
42. An, K.; Song, K.B.; Hur, K. Incorporating charging/discharging strategy of electric vehicles into security-constrained optimal power flow to support high renewable penetration. *Energies* **2017**, *10*, 729. [CrossRef]

applied sciences

MDPI

Article

Application of a DC Distribution System in Korea: A Case Study of the LVDC Project

Juyong Kim *, Hyunmin Kim, Youngpyo Cho, Hongjoo Kim and Jintae Cho

Smart Power Distribution Lab., KEPCO Research Institute, 105 Munji-Ro, Yuseong-Gu, Daejeon 34056, Korea; hm.kim@kepco.co.kr (H.K.); yp.zo@kepco.ko.kr (Y.C.); hongjoo.kim@kepco.co.kr (H.K.); jintae.cho@kepco.co.kr (J.C.)

* Correspondence: juyong.kim@kepco.co.kr; Tel.: +82-42-865-5951

Received: 28 December 2018; Accepted: 6 March 2019; Published: 14 March 2019

Abstract: With the rapid expansion of renewable energy and digital devices, there is a need for direct current (DC) distribution technology that can increase energy efficiency. As a result, DC distribution research is actively underway to cope with the sudden digitization and decentralization of load environment and power supply. To verify the possibility of DC distribution, Korea Electric Power Corporation (KEPCO) Research Institute made a DC distribution system connected with a real power system in Gwangju. The construction of the demonstration area mainly includes design of protection and grounding systems, operating procedures of insulation monitoring device (IMD), and construction of power converters. Furthermore, this paper goes beyond the simulation and the lab testing to apply DC distribution to a real system operation in advance. It is designed as a long-distance low-loaded customer for rural areas and operated by the DC distribution. In addition, safety and reliability are confirmed through field tests of DC distribution elements such as power conversion devices, protection and grounding systems. In particular, to improve the reliability of non-grounding system, the insulation monitoring device was installed and the algorithms of its operational procedures are proposed. Finally, this paper analyzes the problems caused by operating the actual DC distribution and suggests solutions accordingly.

Keywords: DC distribution system; AC/DC converter; protection; grounding system; insulation monitoring device (IMD)

1. Introduction

The share of DC power consumption in PCs, TVs, DC buildings, Internet Data Centers (IDCs), and DC homes is expected to increase. In particular, EPRI in the United States estimates that digital devices will account for 50% of the world's total DC load in 2020. In addition, due to the expansion of renewable energy, such as photovoltaics (PV) generation and fuel cells, there is a need for a new high-quality electricity service, such as DC distribution service technology. In existing AC distribution systems, the power conversion to DC was required to take advantage of the distributed resources or the DC consumers, which resulted in additional cost and power losses. However, in the case of DC distribution systems, the power conversion step can be omitted to enable a more efficient and economical operation, compared to the existing AC distribution system, when the DC distributed resources and DC consumers are connected [1]. Previous studies have confirmed the economics of the Direct Current (DC) distribution system [2]. In addition, the DC distribution system can effectively control voltage using the power conversion device instead of the transformer, and it can limit the fault current in case of an accident at the load side and prevent accidents from spreading to the main grid. However, there are still difficulties for DC distribution system operation and there is insufficient research on connecting with the actual grid system. The accurate detection and location of fault results in fast restoration of the system are needed [3]. In previous studies, the lab test was performed prior to applying the DC distribution to the actual grid system, confirming the tremendous potential of the DC distribution system [4]. In this paper, the DC distribution was applied to the actual grid system and confirmed the possibility of its commercialization. The construction of a demonstration area has been verified by simulations and field tests. Also IMD, the DC distribution fault detecting system, has been installed and its operating algorithms are proposed.

2. Design of DC Distribution System

This paper represents the design of the DC distribution system for the long-distance low load in rural area. Because the DC distribution system is entirely different from the existing AC distribution system, it should begin with the fundamental reviews of the components to supply DC power to the customers [5–7]. This chapter represents the design of DC distribution systems, including protection and grounding.

2.1. Grounding and Protection

The grounding and protection of the DC distribution system in constructing the DC distribution system are an indispensable factor for the stable operation of the system, together with the safety of the users and the protection of the equipment. This chapter describes grounding and protection methods to be applied to the DC distribution system.

2.1.1. Grounding

The classification of grounding, which saves a human life and protects properties from electric shock that could cause fire and lightning, follows the protection method of IEC 60364, IT(Isolate-Terra), TN(Terra-Neutral), and TT(Terra-Terra) methods. Each grounding method is classified in accordance with the purpose and the use. In this study, the IT grounding system, which is deemed to be the most suitable for the DC distribution system, is applied to the plan. In the case of the IT grounding method, none of the power lines are grounded to earth, and only the consumer side conductive enclosure is grounded. Unlike other grounding methods, the IT grounding method is suitable for the DC distribution system because it does not cause any electrolytic corrosion appearing on the earth pole [8].

In order to check the safety of the IT grounding system, the ground fault simulation, which is DC distribution with a 750 Vdc of 1 km line and IT grounding system, was performed (Figure 1). When a ground fault is simulated on the pole, no-fault current flows and the rise of ground potential is zero. In practice, however, a slight voltage rise may occur due to the isolation between the system and the ground. In the case of IT grounding, the ground potential rise does not occur when a ground fault occurs. However, when the second ground fault occurs, a fault current flows to the ground, and the rise of ground potential is very steep. According to reference [9], the rise of ground potential is less than 10 mV when the insulation resistance is 1 MΩ, and it is considered as a ground fault when the insulation resistance value is less than 20 Ω. Figure 2 shows the relation between the pole voltage, insulation resistance, and touch voltage of the DC distribution system of the IT ground. As can be seen from the figure, the IT grounding system hardly causes the ground potential rise, even if the insulation level is low. The ground fault in the underground and overhead lines are then simulated. The graph shown in Figure 3a shows the simulation result of the contact voltage at the ground fault in the underground line. In the event of a fault, a voltage close to the terminal voltage is generated and is decreased in 4 μs, falling below the maximum contact voltage of 120 Vdc. After 10 μs, the contact voltage drops below 1 Vdc. Since the period of a dangerous level of voltage generation is very short, the danger to the human body is very low.

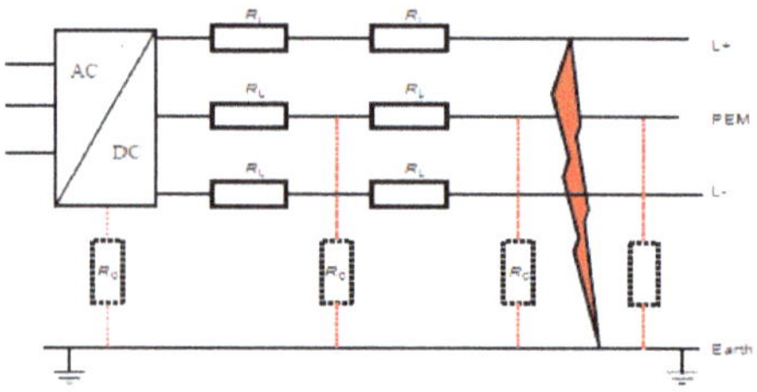

Figure 1. Ground Fault IT(Isolate-Terra) System.

Figure 3b shows the ground fault in the overhead line. In the case of an overhead line, the resonance of the contact voltage occurs at the beginning of the fault due to the inductance of the line. Theoretically, the maximum impulse voltage should be the same as the pole voltage at the beginning of the fault. However, the simulation results show that the initial impulse voltage is relatively low because of the low capacitance of the line. However, in the case of the overhead line, in contrast to the underground line, the resonance phenomenon of contact voltage attributed to fault occurred for a relatively long 100 μs. However, the voltage exceeding 120 Vdc dissipates almost instantaneously with the occurrence of the fault. Therefore, one can conclude that safety is not a concern in this case. Based on international standards, it is acceptable for the human body to be exposed to 800 Vac for 0.04 s. The simulation results of the aforementioned overhead line and the underground line all disappeared within 100 μs. Therefore, it is considered that the discharge current due to the total capacitance between the DC line and the ground in the IT system moves quickly, and no danger to the human body is anticipated. Given that the TT and TN grounding methods connect one power line to the ground through a protective conductor, a ground fault in any one of the power lines causes a fault current to be generated because of the closed circuit between the power sources by the protective conductor. The IT grounding method, on the other hand, is advantageous even in the case that the one-wire ground fault in the power line does not cause an interruption of the power or the load supply because it does not constitute a closed circuit. However, even when the IT grounding method is used, if a two-wire ground fault occurs, a closed circuit is formed and causes a ground fault. Therefore, in order to secure the advantages of the IT grounding method and prevent electric shock accidents, it is important to detect the accident on the one-wire ground fault and take measures to prevent a two-wire ground fault. The IT grounding method is characterized by a decrease in the insulation resistance between the power line and the ground in the event of a ground fault. Such deterioration of insulation resistance leads to insulation breakdown, which can cause damage due to electric shock and fire. To prevent such failure, the IEC 61557 specifies that an insulation monitoring device (IMD) should be installed in an IT grounding system environment to protect facilities and the human body from one-line ground fault [10]. Therefore, an IMD is installed and applied in this study.

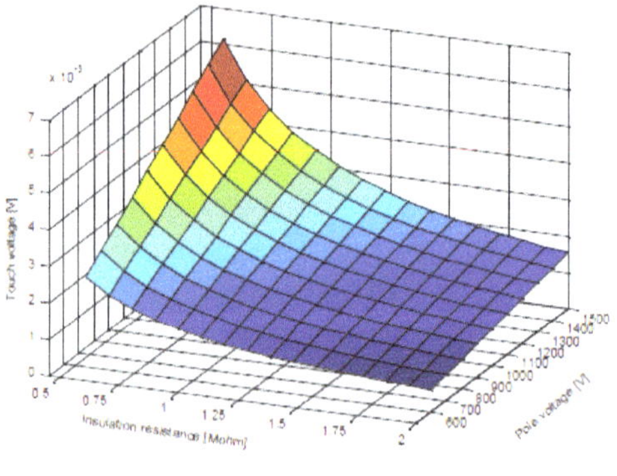

Figure 2. Correlation diagram of pole voltage, insulation level, and contact voltage in IT ground.

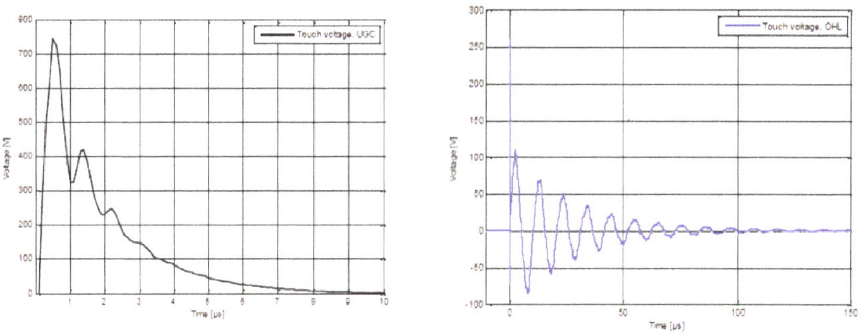

(**a**) (**b**)

Figure 3. Simulation of ground fault in DC distribution line of IT ground: (**a**) underground line; (**b**) overhead line.

2.1.2. Fault Detection Device

In the case of non-grounded systems, it is difficult to detect grounding faults by a conventional method. Therefore, real-time analysis of the insulation resistance in the energized-line is necessary. Thus, an IMD is installed.

IMD is commonly used to measure insulation resistance in closed circuit rail or medical devices [11]. Currently, the development of IMD technology for the distribution line is in the early stage, and there is no application for outdoor use. Furthermore, it is necessary to study operation standards and procedures. Therefore, this paper confirms the applicability of IMD and presents the operational procedures. It is necessary to set appropriate alarm value to detect ground faults while operating the distribution system and coping with system stoppage or re-input according to the situation. However, there is currently no standard for IMD alarm set-points for DC distribution lines. However, in "Selection and erection of electrical equipment—isolation, switching and control" of IEC 60364-5-53 Annex H, the pre-warning is specified as 100 Ω per 1 V and the warning as 50 Ω per 1 V. The setting value of IMD is recommended to be set to 1.5 times this value. It is essential to select an appropriate alarm set-point because the IMD for the DC distribution line monitors the insulation of the DC distribution line and the associated power converter. Therefore, a flow chart of the operation procedure was proposed for determining the IMD alarm setting value required in the field application (refer to Figure 4). In the case of alarm 1 of IMD, be cautious of the possibility of decreased insulation resistance in the line, while the alarm 2 should immediately trigger stopping of the operation, followed by an inspection. This is to prevent possible grounding caused by animals or humans, which can lead to secondary failure, which may lead to deadly accidents. In this study, the alarm value is set by measuring the real-time insulation resistance value according to the proposed operating procedure for the DC distribution to be applied to the actual grid system.

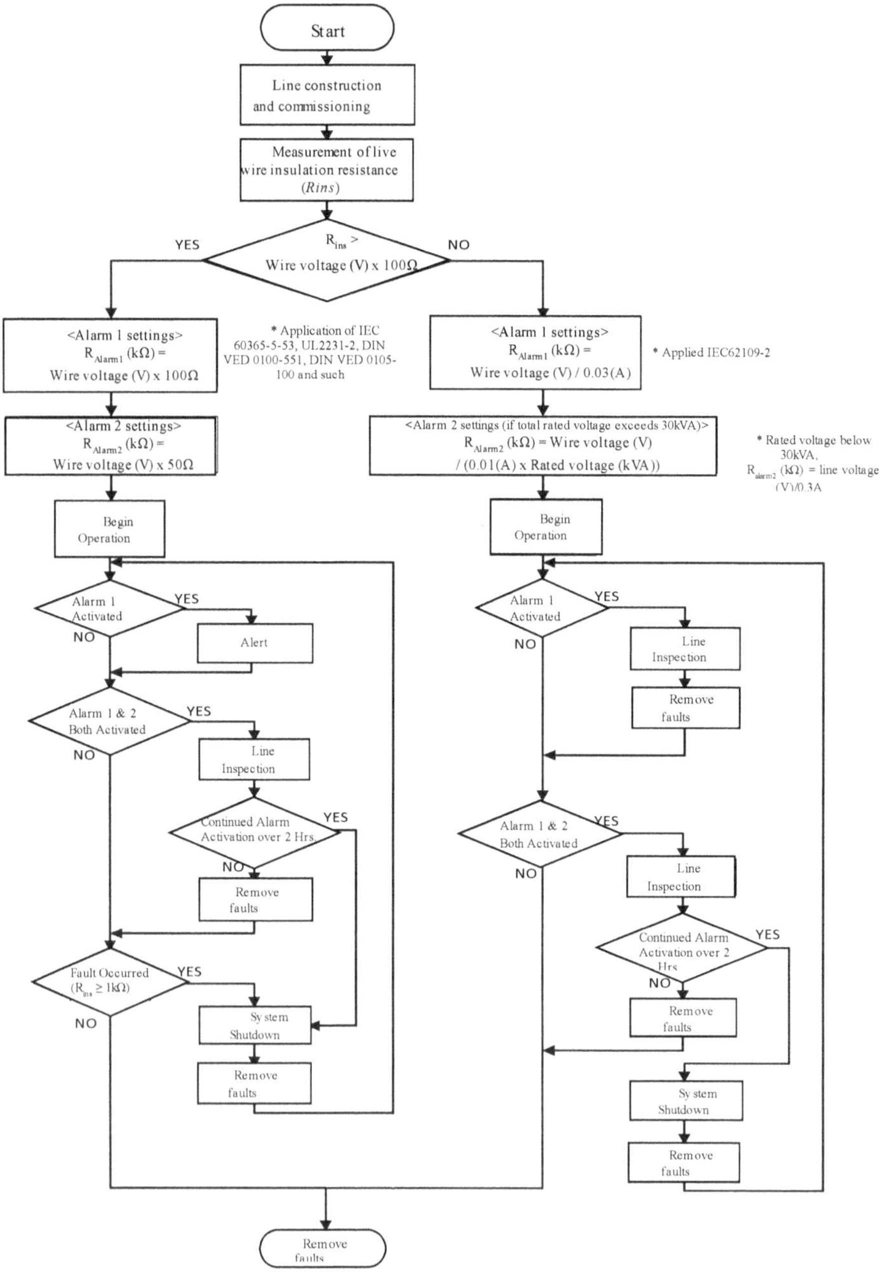

Figure 4. Proposed insulation monitoring device (IMD) operating procedures.

2.1.3. Protection

The purpose of the power system protection is to secure people and equipment by eliminating faults or limiting voltage and current below a certain level. If a failure occurs in the DC distribution system, the protection operation should be performed for the faulty part, and the fault should be detected, even when it does not require any protective operation [12,13].

In the DC distribution system, the protection system can be divided into three parts: power supply rectifier, DC distribution line, and consumer. The protection system should be determined by the system grounding method, voltage level, network configuration, and power converter structure. Protection of the system shall meet the following requirements.

- The risk level should not be higher than the existing AC system.
- Faults should be detected and removed.
- Fault section should be separated.
- There should be no breakdown of equipment due to line failure.
- Safety should not be threatened by equipment failure.

The protection requirements of the above system can be satisfied by protective devices such as converter protection system, circuit breaker (CB), overcurrent/overvoltage relay, surge protection device (SPD), IMD, grounding system, electrical isolation, and insulation cooperation.

As shown in the Figure 5, the protection system of the DC distribution system was designed. First of all, a surge protection device needs to be installed on the primary side of the transformer to protect against overvoltage which can flow through the high-voltage line. The transformer does not need to be grounded on the secondary side because the flow of current through the earth can cause permanent failure. A DC circuit breaker (DCCB) was installed on the rectifier output side to prepare for overcurrent and short circuit faults that may occur in each pole. In addition, the DCCB can be used when the rectifier between the high voltage line and the DC line needs to be independent, allowing for the independent operation of the lines on each end. At this moment, the converter system can supply limited fault current when the fault occurs and control the DCCB by the measured fault results. Since the use of the IT grounding system, the IMD is installed to monitor grounding faults. If a grounding fault occurs in the DC line, the insulation monitoring device controls the DCCB, and if the grounding situation is not resolved, the DCCB between the transformer and the converter system is controlled to cut off the power supply. In addition, an SPD is installed to protect the converter system and the line against the DC line overvoltage input. Likewise, an SPD is installed in front of the customer's side to provide protection against overvoltage inflow to the customer's network and the inverter. In addition, a DCCB is installed in front of the customer's inverter to protect the DC line from the fault of the customer's network and the inverter. The inverter limits the amount of current passing through at the time of the fault to ensure the normal operation of the CB, while the CB installed on the customer's power supply line performs the protection operation if the operating characteristics curve in the event of failure.

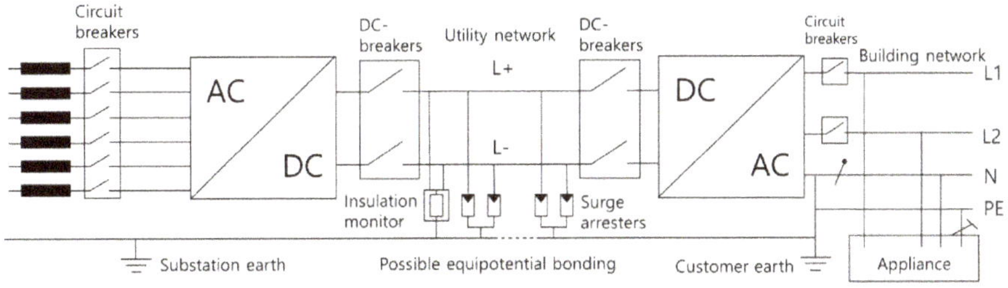

Figure 5. Design of protection system for DC distribution.

3. Field Test

Before applying the DC distribution element to the field, the application test of the IMD and the converter system device is carried out on the DC distribution line installed in the power test center.

The power test center is a demonstration site similar to the environment in which actual grid system are installed. The center was constructed to improve the reliability of a device under test (DUT) and to confirm its performance (Figure 6 and Table 1). A 70SQ CV cable on the demonstration site, a long-range 700 m power line, was used in the test.

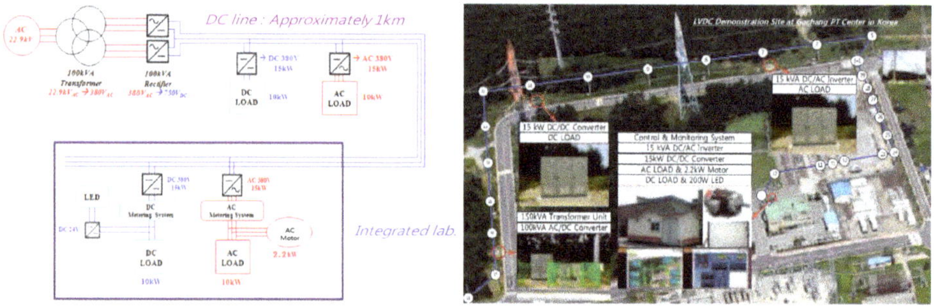

Figure 6. Configuration of DC demonstration site.

Table 1. Specifications of component in power test center.

Item	Capacity	Topology
AC/DC Converter (Rectifier)	100 kW	2-level type
DC/DC Converter (2EA)	15 kW	Buck-Boost type
DC/AC Converter (2EA)	15 kW	2-level type
Transformer	100 kVA	-

3.1. Fault Detection

To verify the IMD, at the DC demonstration site, the application test was divided into two cases.

(a) Case 1

The condition of first IMD application test is shown in Figures 7 and 8. As shown in the figures, the P-pole grounding fault occurred in the middle of the power line, and the measurement position of the IMD is the output side of the AC/DC converter. Also, the short circuit resistance of the grounding fault simulator was set to 0.632 Ω to limit the grounding fault current to 600 A maximum, and the fault duration was set to 10 s. The IMD generated Alarm 1 and 2 at approximately 5 s after the fault occurred.

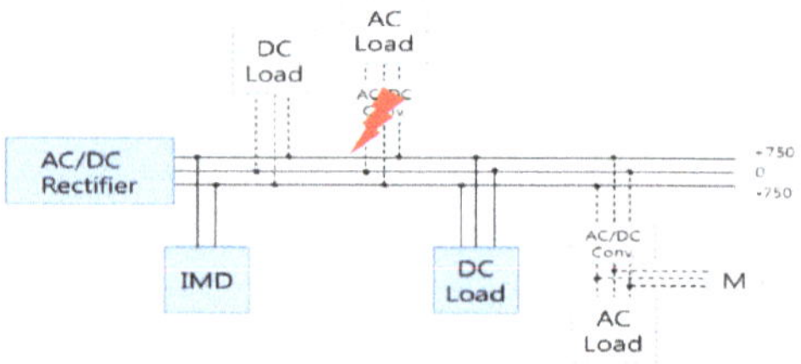

Figure 7. Fault detection test case 1.

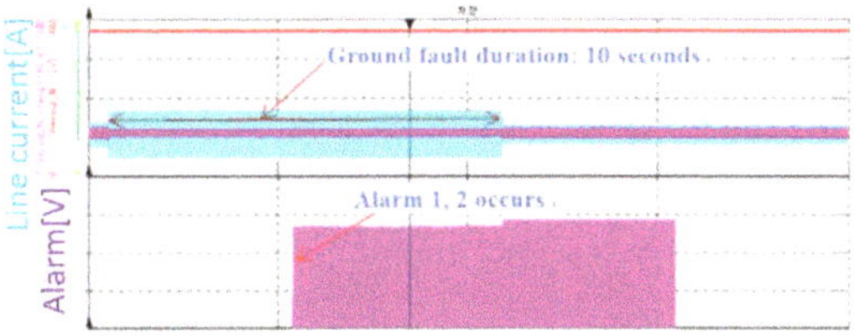

Figure 8. Fault detection test case 1 result waveform.

(b) Case 2

In case 2, the IMD performance test was conducted with varying loads. The configuration is shown in Figure 9. The real-time insulation resistance of the IMD was measured when the IMD was connected to the AC/DC converter output side, and each load of the demonstration site was driven. The insulation resistance of the IMD was measured in real time while the IMD was connected to the output side of the converter to pass the voltage through the line, and all the loads at the demonstration site were applied. Table 2a shows the measured values when the converter is activated in the no-load condition.

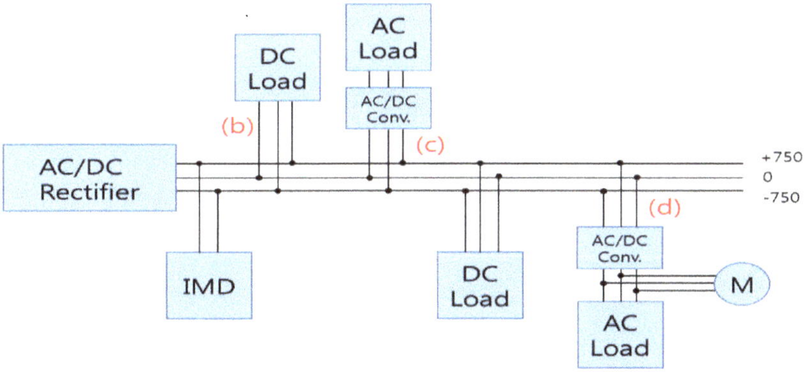

Figure 9. Fault detection test case 2.

The real-time insulation resistance value is above the maximum measurable value of 10 MΩ. Case (b)–(d) show the real-time insulation resistance value when the simulated loads corresponding to Figure 9. It shows a healthy level of insulation resistance of the test line and power converters from 225 kΩ to 295 kΩ. Therefore, it was confirmed that the distance and variation of load do not significantly impact the insulation resistance value. Also, it was found that the failure detection time and the failure resolution recognition time were about 5 s, therefore the instantaneous response was impossible.

Table 2. Measurement of insulation resistance.

Case	Insulation Resistance
(a)	>10 MΩ
(b)	224 kΩ
(c)	253 kΩ
(d)	295 kΩ

4. Construction of Actual DC Distribution Line

In this chapter, applying actual DC distribution studies carried out based on the previous studies are discussed. This chapter established DC distribution system to connect to the actual grid system and confirmed the feasibility of the DC distribution system.

4.1. Commercial Distribution Line

The existing long-distance low-load grid line, which supplied high-voltage AC power, is replaced by the DC distribution system. The target line is a long-distance (over 12 spans) and low-load (under 30 kW) aerial bundled cable (ABC), where the possibility of line failure due to contact with trees is high. The location of the target line is 52L1~52L12 at Chungok Reservoir, a Gwangju Jeonnam HQ controlled area.

In, DC distribution, a DC/AC inverter has been used to convert a 13.2 kV high-voltage AC line into DC line. Unlike any conventional transformer, controlling voltage and current at the output stage is possible, as well as the power factor through the internal control algorithm of the converter device. In addition, since the low-voltage line replaces the existing high-voltage line, it is expected that the fault that occurs due to the contact of the trees can be reduced.

4.2. DC Distribution Equipment

The significant equipment of the DC distribution system connected to the actual grid in Gwangju are primarily composed of the transformer, converter system, protection system, and monitoring system. The specifications of each device are shown in Table 3. Single-phase AC/DC converter and single-phase DC/AC inverter switching devices use SiC for efficiency improvement [14,15].

Table 3. Specifications of DC distribution equipment.

Item	Pole Transformer	Converter (AC/DC)	Inverter (DC/AC)
Input voltage	1∅ 13.2 kVac	1∅ 220 Vac	750 V_{dc}
Capacity	30 kVA	30 kW	30 kW
Rated voltage	-	750 V_{dc}	220 Vac ± 5%
THD	-	<5%	<5%
Efficiency	-	≥97%	≥97%

The detailed system configuration is shown in Figure 10. In case of a problem with the DC distribution system, the customer can switch to the existing AC system in order to prepare for the long-term power outage. The Cut-Out Switch (COS) shown in Figure 10 and the pole transformer on the load side can supply AC. To supply DC voltage, open the COS (①) for the two lines and put COS (②) to use. On the contrary, to supply AC voltage, put COS (①) to use for the two lines and open the COS (②). The power supply and the load side AC/DC Converter are separated from the transformer to ensure stability. A low-voltage line was constructed below the neutral line in consideration of the induction voltage of the high-voltage line. Also, to build a non-grounding (IT) system, the grounding of the enclosure through the three-class grounding was enabled, and the IMD for monitoring the grounding fault of the line was installed.

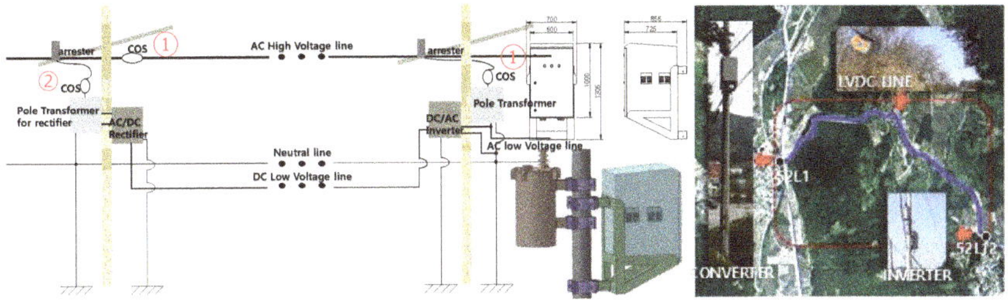

Figure 10. Actual DC distribution line configuration.

4.3. Operation Results

The DC distribution was applied to the actual gird line to find problems and to improve the performance. Also, by establishing operating data for the pilot lines, design, and operation standards for DC distribution lines were created.

The performance of the AC/DC converter applied to the actual grid line was verified. The average daily data from 14 February 2017 to 30 July 2017 are summarized. Approximately 86,000 data points were collected per day, and the average was calculated. The voltage was very stable, with a fluctuation of 5% over the 6 months. Except for the 23 February 2017 the efficiency remained near 98%.

On 23 February 2017 because the load usage was zero from 06:37:54 to 10:09:14, the daily average efficiency was low. The data showed that the performance of the AC/DC converter system was satisfactory.

As shown in the Section 3.1 "Fault detection," the IMD was installed on the output side of the AC/DC Converter as the IMD installation location does not impact its performance. In particular, this paper tested the feasibility of the outdoor operation of IMD by installing it in an outdoor environment, which is rarely done. The insulation resistance value (R_{INS}) of the energized-line was measured upon the installation of the IMD, following the operation manual shown in Figure 4. The measured value was 487 kΩ so R_{INS} is 750 (line voltage) × 100 Ω × 1.5 = 112.5 kΩ or more. Therefore, the alarm set value is set as follows.

(i) Alarm 1—pre-warning
R_Alarm1 (kΩ) = line voltage (V) × 100 (Ω) × 1.5 = 750 × 100 × 1.5 = 112.5 (kΩ)

(ii) Alarm 2—warning
R_Alarm2 (kΩ) = line voltage (V] × 50 (Ω) × 1.5 = 750 × 50 × 1.5 = 56.25 (kΩ)

Next, Figure 11 is a measure of the IMD resistance value obtained from the actual DC distribution system. The average daily data from 14 February 2017 to 30 July 2017 are summarized. As can be seen from the waveform, the insulation resistance value is maintained at approximately 486 kΩ during the measurement period, indicating alarms and faults did not occur. It was confirmed that the insulation resistance value was monitored in real time.

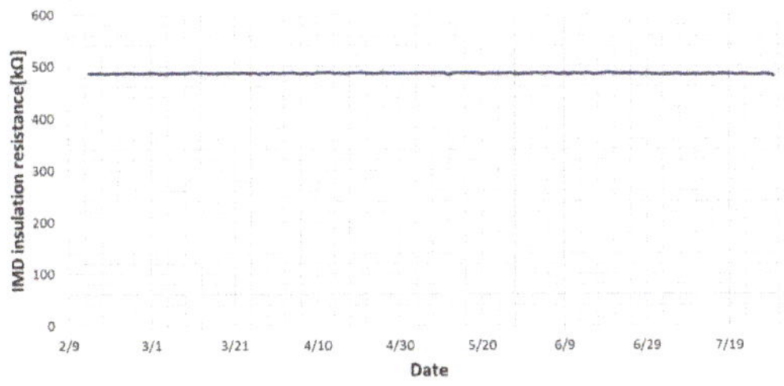

Figure 11. Measurement of insulation resistance.

The actual DC distribution system was completed in October of 2016 and continues to operate normally. The waveforms below in Figure 12 shows the operation data of the AC/DC Converter (input/output voltage, power) from 22 February 2017. As the load usage fluctuation is constant, a representative day's waveform was selected to be analyzed. The load usage ranged between 3 kW and 4 kW, with a sudden maximum peak at 9 kW. Although the load was not large, the protection operation of the power conversion device occurred intermittently in a sudden change of load condition that was not periodic.

The representative waveform at the time of the protection operation is shown in Figure 13. It is evident that the operation of the converter was interrupted at the moment when the protection operation was initiated. To understand this deeply, the current data at the time of the incident was acquired and analyzed with increased sampling time.

As shown in the waveform in Figure 13, 121 A of output current triggered the protection system. This exceeds the rated capacity of the converter and the threshold current of triggering protection system (Maximum 60 A) that indicates the need for more analysis.

A field survey confirmed that frequent water pump motor operations were found at farmhouses. The motor inrush current is commonly known to be 500%~700% of the rated current. However, in this incident, the inrush current reached 10 times the rated current. The DC/AC inverter system improvement plan is proposed as follows.

There was no inrush current problem in the operation of the DC distribution line since the protection operation level of the converter was reset. KEPCO Research Institute will continue to develop new problems and solutions through a distribution operation.

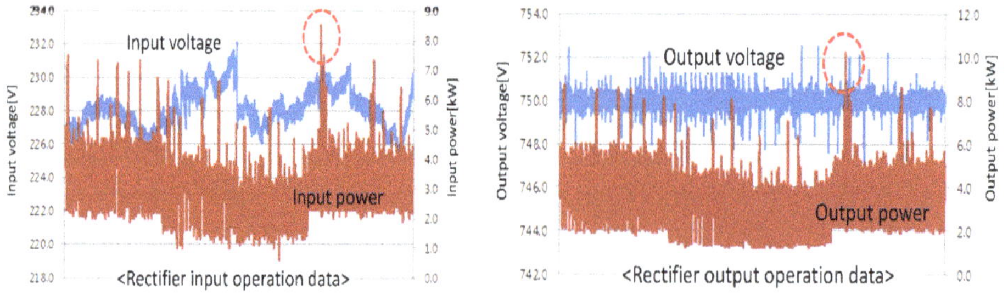

(a) (b)

Figure 12. Input and output data of AC/DC Converter. (**a**) Input voltage and power. (**b**) Output voltage and power.

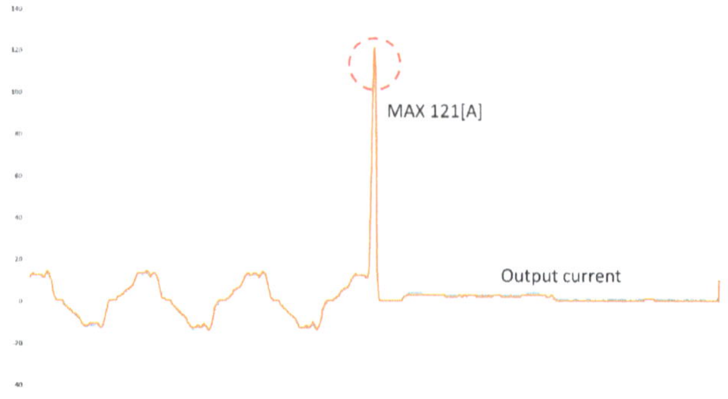

Figure 13. Waveform of protective action.

5. Conclusions

Based on the business model of the DC distribution system derived from the previous work, KEPCO research institute confirmed the possibility by applying DC distribution to the actual grid system. The power line subject to the test was set to Section 52L1–52L12 at Chungok Reservoir, a Gwangju Jeonnam HQ controlled area with the ABC cables. The target line is a long-distance low-load grid rural area with 12 spans or more. A 30 kW class AC/DC converter for supplying DC voltage and a 30 kW class DC/AC inverter for supplying AC voltage for household use were demonstrated to construct a DC distribution in the target line. In addition, an IT grounding system that is safe for electrolytic corrosion is applied to the DC distribution, and an IMD was installed to detect the failure situation in real time. Since the IMD is rarely used in distribution lines in general, KEPCO proposed the operational procedures and verified the feasibility of IMD by carrying out demonstration site tests accordingly. In addition to adopting the IT grounding system, the safety system of the DC distribution system was secured by designing and constructing the protection system.

By applying DC distribution to the actual grid system, the possibility of replacing the AC high voltage line to solve the operational problems such as contact with trees and to operate the stable line was confirmed. The established DC distribution is currently in operation. The operation data is and will be continuously analyzed to improve the deficiencies. In the future, KEPCO plans to construct an DC distribution system, with DC consumers, that will be more efficient and self-sustaining with connected distributed power resources.

Author Contributions: The following statements should be used "Conceptualization, J.K.; Methodology, J.K.; Software, H.K. (Hongjoo Kim); Validation, Youngpyo Cho, H.K. (Hyunmin Kim); Formal Analysis, J.C.; Investigation, Y.C.; Resources, H.K. (Hyunmin Kim); Data Curation, Y.C.; Writing-Original Draft Preparation, H.K. (Hyunmin Kim); Writin g-Review & Editing, J.K.; Visualization, J.C.; Supervision, J.K.; Project Administration, J.K.

Funding: This research was funded by Korea Electric Power Corporation (KEPCO) Research Institute grant number R15DA12.

Conflicts of Interest: The authors declare no conflicts of interest.

References

1. Yahoui, H.; Vu, H.G.; Tran, T.K. A control strategy for DC Smart Grids operation. In Proceedings of the 2019 First International Symposium on Instrumentation, Control, Artificial Intelligence, and Robotics (ICA-SYMP), Bangkok, Thailand, 16–18 January 2019.
2. Afamefuna, D.; Chung, I.Y.; Hur, D.; Kim, J.Y.; Cho, J. A Techno-Economic Feasibility Analysis on LVDC Distribution System for Rural Electrification in South Korea. *J. Electr. Eng. Technol.* **2014**, *9*, 1501–1510. [CrossRef]
3. Sharanya, M.; Devi, M.M.; Geethanjali, M. Fault Detection and Location in DC Microgrid. In Proceedings of the 2018 National Power Engineering Conference (NPEC), Madurai, India, 9–10 March 2018.
4. Cho, J.; Kim, J.H.; Chae, W.; Lee, H.J.; Kim, J. Design and Construction of Korean LVDC Distribution System for supplying DC Power to Customer. In Proceedings of the CIRED Conference on 23th International Conference on Electricity Distribution, Lyon, France, 15–18 June 2015; pp. 1–4.
5. IEC 60364-4-41. *Low Voltage Electrical Installations—Part 4-41: Protection for Safety—Protection Against Electric Shock*; International Standard; Fifth edition 2005-12; IEC: Geneva, Switzerland, 2005.
6. IEC 60364-4-44. *Low Voltage Electrical Installations—Part 4-44: Protection for Safety—Protection Against Voltage Disturbances and Electromagnetic Disturbance*; International Standard; Edition 2.0 2007-08; IEC: Geneva, Switzerland, 2007.
7. Karppanen, J.; Kaipia, T.; Mattsson, A.; Lana, A.; Nuutinen, P.; Pinomaa, A.; Peltoniemi, P.; Partanen, J.; Cho, J.; Kim, J.; et al. Selection of Voltage Level in Low Voltage DC Utility Distribution System. In Proceedings of the CIRED Conference on 23th International Conference on Electricity Distribution, Lyon, France, 15–18 June 2015; p. 1174.

8. Hirose, K.; Tanaka, T.; Babasaki, T.; Person, S.; Foucault, O.; Sonnenberg, B.J.; Szpek, M. Grounding concept considerations and recommendations for 400VDC distribution system. In Proceedings of the Telecommunications Energy Conference (INTELEC), Amsterdam, The Netherlands, 9–13 October 2011; pp. 1–8.
9. Salonen, P.; Kaipia, T.; Nuutinen, P.; Peltoniemi, P.; Partanen, J. A Study of LVDC Distribution System Grounding. In Proceedings of the Conference NORDAC, Bergen City, Norway, 8–9 September 2008.
10. Jiang, J.; Ji, H. Study of insulation monitoring device for DC system based on multi-switch combination. In Proceedings of the 2009 Second International Symposium on Computational Intelligence and Design, Changsha, China, 12–14 December 2009; Volume 1, pp. 429–433.
11. Wolfgang Hofheinz. *Protective Measures with Insulation Monitoring—Application of Unearthed IT Power Systems in Industry, Mining, Railways, Marine/Oil and Electric/Rail Vehicles*, 3rd ed.; VDE Verlag: Berlin, Germany, 2006.
12. Salonen, P.; Nuutinen, P.; Peltoniemi, P.; Partanen, J. Protection scheme for an LVDC distribution system. In Proceedings of the 20th International Conference and Exhibition on Electricity Distribution (CIRED 2009), Prague, Czech Republic, 8–11 June 2009.
13. Salomonsson, D.; Soder, L.; Sannino, A. Protection of low-voltage DC microgrids. *IEEE Trans. Power Deliv.* **2009**, *24*, 1045–1053. [CrossRef]
14. Naakka, V. Reliability and Economy Analysis of the LVDC Distribution System. Master's Thesis, Tampere University of Technology, Tampere, Finland, 2012.
15. Verma, A. Algorithm for Design of Digital Notch Filter Using Simulation. *Int. J. Adv. Res. Artif. Intell.* **2013**, *2*, 40–43. [CrossRef]

applied sciences

MDPI

Article

A Quantitative Index to Evaluate the Commutation Failure Probability of LCC-HVDC with a Synchronous Condenser

Jiangbo Sha, Chunyi Guo *, Atiq Ur Rehman and Chengyong Zhao

State Key Laboratory of Alternate Electrical Power System with Renewable Energy Sources, North China Electric Power University, Beijing 102206, China; m18810911471@163.com (J.S.); atiq_marwat99@yahoo.com (A.U.R.); chengyongzhao2@163.com (C.Z.)

* Correspondence: chunyiguo@gmail.com; Tel.: +86-15911058467

Received: 9 January 2019; Accepted: 27 February 2019; Published: 5 March 2019

Featured Application: The new index of area ratio of the commutation failure probability is expected to evaluate the commutation failure under a wider fault range from the transient performance, which is helpful to evaluate the impact of dynamic reactive power compensators on the commutation failure probability of LCC-HVDC.

Abstract: Since thyristor cannot turn off automatically, line commutated converter based high voltage direct current (LCC-HVDC) will inevitably fail to commutate and therefore auxiliary controls or voltage control devices are needed to improve the commutation failure immunity of the LCC-HVDC system. The voltage control device, a synchronous condenser (SC), can effectively suppress the commutation failure of the LCC-HVDC system. However, there is a need for a proper evaluation index that can quantitatively assess the ability of the LCC-HVDC system to resist the occurrence of commutation failures. At present, the main quantitative evaluation indicators include the commutation failure immunity index and the commutation failure probability index. Although they can reflect the resistance of the LCC-HVDC system to commutation failures to a certain extent, they are all based on specific working conditions and cannot comprehensively evaluate the impact of SCs on suppressing the commutation failure of the LCC-HVDC system under certain fault ranges. In order to more comprehensively and quantitatively evaluate the influence of SCs on the commutation failure susceptibility of the LCC-HVDC system under certain fault ranges, this paper proposes the area ratio of commutation failure probability. The accuracy of this new index was verified through the PSCAD/EMTDC. Based on the CIGRE benchmark model, the effects of different synchronous condensers on LCC-HVDC commutation failure were analyzed. The results showed that the new index could effectively and more precisely evaluate the effect of SCs on commutation failures. Moreover, the proposed index could provide a theoretical basis for the capacity allocation of SCs in practical projects and it could also be utilized for evaluating the impact of other dynamic reactive power compensators on the commutation failure probability of the LCC-HVDC system under certain fault ranges.

Keywords: commutation failure probability; line commutated converter; high voltage direct current (HVDC); synchronous condenser (SC); quantitative evaluation

1. Introduction

The LCC-HVDC system is widely used in the world due to its numerous advantages such as bulk direct current (DC) power transmission over long distances and asynchronous interconnection of alternating current (AC) grids [1]. However, due to the use of thyristor technology, the LCC-HVDC system can exhibit the commutation failure during various system disturbances [2]. One of the main reasons for commutation failure is the AC busbar voltage drop that occurs during fault conditions [3,4]. Therefore, it is necessary to quickly and effectively regulate the AC busbar voltage during faulty conditions so that the probability of commutation failure can be reduced. Dynamic reactive power compensators can be installed at the inverter side to achieve quick voltage regulation of the AC busbar by supplying a surplus of reactive power and thus decreasing the probability of commutation failure [5]. The synchronous condenser (SC), a type of dynamic reactive power compensator, can increase the AC system's strength due to the shunting effect of the transient reactance during steady and transient states [6]. It is therefore widely used in real-world LCC-HVDC systems to effectively regulate the AC voltage and thus avoid commutation failure. Recently, SCs have been installed in several LCC-HVDC projects in China to enhance the system's strength and mitigate commutation failures. Hence it is necessary to present an index that can quantitatively and comprehensively assess the influence of SCs on the commutation failure of the LCC-HVDC systems.

Several works have focused on evaluating the commutation failure of the LCC-HVDC system. An area of vulnerability (AOV) was discussed and a mechanism for the optimal allocation of dynamic VAR sources was proposed in [7] in order to decrease the commutation failure probability of the LCC-HVDC system at the inverter end. However, due to the use of steady-state equations for calculating the probability of commutation failure, the transient behavior of the extinction angle and phase shift in the AC voltage were not considered, which could lead to a decrease in the accuracy of the results. An analytical approach to investigate the immunity levels to commutation failure of the multi-infeed LCC-HVDC system was presented [8]. However, the dynamic responses during transient conditions and the effects of different reactive power compensators were not considered. A fast calculation approach to evaluate the risk of commutation failure in the multi-infeed LCC-HVDC system was proposed [9]. However, the transient responses of different control parameters were not undertaken. The commutation failure in the LCC-HVDC system using the spatial-temporal discreteness of AC disturbances was investigated [10]. However, this approach was based on a steady-state model, which meant that results obtained under different fault situations could have a lack in accuracy. Two indices, referred to as the commutation failure immunity index (CFII) and commutation failure probability index (CFPI), were proposed in [11]. This approach was based on transient simulations, however could only evaluate the commutation failure under certain faulty conditions. Thus, it is meaningful to propose a more comprehensive index to evaluate the commutation failure under a wider fault range from the transient performance.

This paper proposes a novel index termed as area ratio of commutation failure probability, to comprehensively and quantitatively evaluate the commutation failure of the LCC-HVDC system with a SC under certain fault ranges. Considering the CIGRE benchmark model, the LCC-HVDC system with a SC was built in the PSCAD/EMTDC. Different scenarios assuming the number of synchronous condenser units were carried out in order to examine its effect on the area ratio of the commutation failure probability of the LCC-HVDC system. The effectiveness of the proposed index was validated through simulation results under different fault types and fault severity conditions.

The paper is arranged as follows. In Section 2, a detailed configuration of the LCC-HVDC system with a SC is presented. In Section 3, the overall control mechanism for the LCC-HVDC systems and a SC is discussed. In Section 4, a new index termed as area ratio of commutation failure probability is proposed and verified through the PSCAD/EMTDC under different fault conditions. In Section 5, the conclusion is provided.

2. Configuration of LCC-HVDC with a Synchronous Condenser

The configuration of the mono-polar LCC-HVDC system with SCs is shown in Figure 1. The HVDC station is composed of two six-pulse converters connected in series. The SC is connected to the AC bus at the inverter side via a step-up converter transformer. The rated DC power of the LCC-HVDC system is 1000 MW and the rated reactive power of the SC is 100 Mvar.

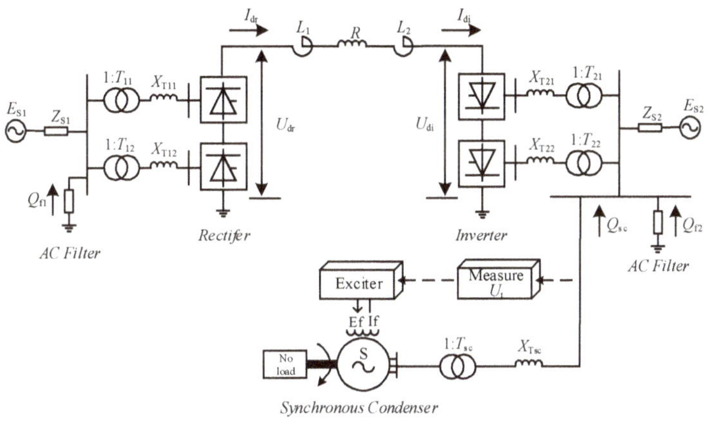

Figure 1. Schematic diagram of the LCC-HVDC system with a synchronous condenser.

The main parameters of the LCC-HVDC system with a SC are summarized in Table 1.

Table 1. Main parameters of LCC-HVDC with a SC.

Parameters	LCC-HVDC	SC
Rated AC voltage/kV	–	20
Rated AC current/kA	–	2.88
Rated capacity/MW(Mvar)	1000	100
DC voltage/kV	500	–
DC current/kA	2	–
Rated ratio of converter transformer/(kV/kV)	345/213.5(Rectifier) 230/209.2(Inverter)	20/230
Short circuit impedance of coupling transformer/%	18(Rectifier) 18(Inverter)	8
Smoothing reactance/H	0.15	–

3. Control Mechanism for LCC-HVDC with a Synchronous Condenser

3.1. Control Mechanism for LCC-HVDC

The control system of the LCC-HVDC is consistent with the CIGRE benchmark model [12]. The system operates in the constant current control mode with a minimum firing angle at the rectifier end, and adopts a constant extinction angle control mode at the inverter end, which is also equipped with a constant current control mode and a current error controller (CEC) mechanism. The constant current control at the inverter side is used as a backup control mechanism during severe AC and DC disturbances. It contains a voltage-dependent current order limiter (VDCOL) function, which helps in limiting the DC overcurrent and avoiding the stress on thyristor valves during different fault conditions. Figure 2 presents the overall control mechanism for the LCC-HVDC system.

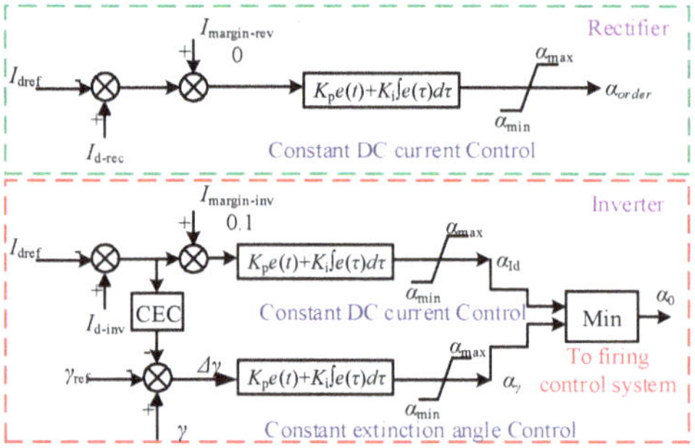

Figure 2. Control mechanism for the LCC-HVDC system.

3.2. Control Mechanism for the Synchronous Condenser

The basic control mechanism for the SC is shown in Figure 3. The measured AC voltage and excitation current, shown as U_t and I_f, respectively, are the input variables, and the excitation voltage E_f is generated via proportional-integral (PI) function and then delivered to the exciter system. The SC adopts the constant voltage control mode to adjust the AC bus voltage during both steady-state and transient situations. When an AC fault is applied at the AC busbar, the increase in the excitation current causes the excitation voltage of the SC to increase in order to provide a certain amount of reactive power needed for AC voltage regulation. If the fault is serious or lasts for a longer time, the excitation current could reach the maximum over-excitation current value. This will activate the maximum excitation current limitation function by the control-signal of the I_f_mark to limit the excitation current around the referenced value of $I_{f\text{max\_ref}}$. However, it is not allowed to keep the maximum over-excitation current as $I_{f\text{max\_ref}}$ for a long time. When the operating time of SC with $I_{f\text{max\_ref}}$ exceeds the allowable time, the anti-time over-excitation limiter will activate the over-excitation current limitation function by the control-signal of OEL_mark, so as to limit the over-excitation current to a level of $I_{f\text{\_ref}}$ under which the SC can keep operating for a longer time, finally to avoid the internal overheating of the SC.

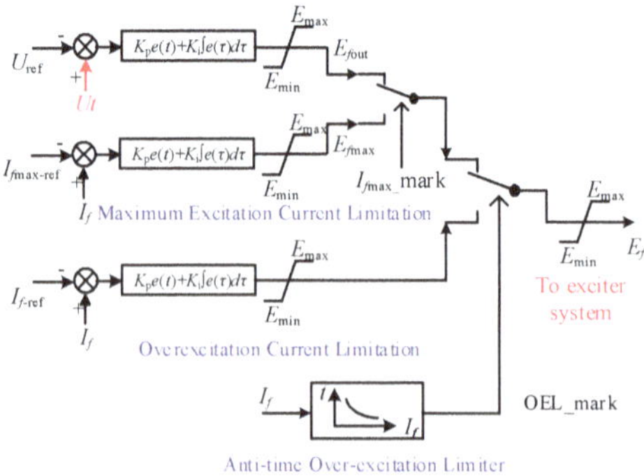

Figure 3. Control mechanism for the synchronous condenser.

4. Using the Area Ratio of Commutation Failure Probability to Evaluate the Effect of SCs on the Commutation Failure Immunity of LCC-HVDC

In order to further study the suppression effect of SCs on the commutation failure of the LCC-HVDC system, firstly, the simulation from PSCAD/EMTDC with different numbers of SCs was compared to verify the effectiveness of the SCs on suppressing the commutation failure. Then, a new index called area ratio of commutation failure probability was proposed. Finally, the area ratio of commutation failure probability index was used to evaluate the suppression effect of the SCs on the commutation failure of the LCC-HVDC system under different short circuit ratios (SCRs) at the inverter side. The following three cases were considered:

***Case 1*:** No synchronous condenser is connected at the AC bus of the inverter side;

***Case 2*:** One synchronous condenser with rated capacity of 100 Mvar is connected at the AC bus of the inverter side;

***Case 3*:** Two synchronous condensers each with rated capacity of 100 Mvar are connected at the AC bus of the inverter side;

4.1. Validation of the Effectiveness of the SCs in Suppressing the Commutation Failure of the LCC-HVDC System

Three-phase fault is the most serious fault among all types of AC faults. Therefore, the same three-phase fault was applied at the inverter bus of three cases to verify the effectiveness of the SC on suppressing commutation failure. The fault inductance value was set to 1.28H (the smallest inductance value that does not result in commutation failure in case 1), the fault time was 5.1s, and the fault duration was 0.05s. The comparison of dynamic responses under the three-phase fault is shown in Figure 4.

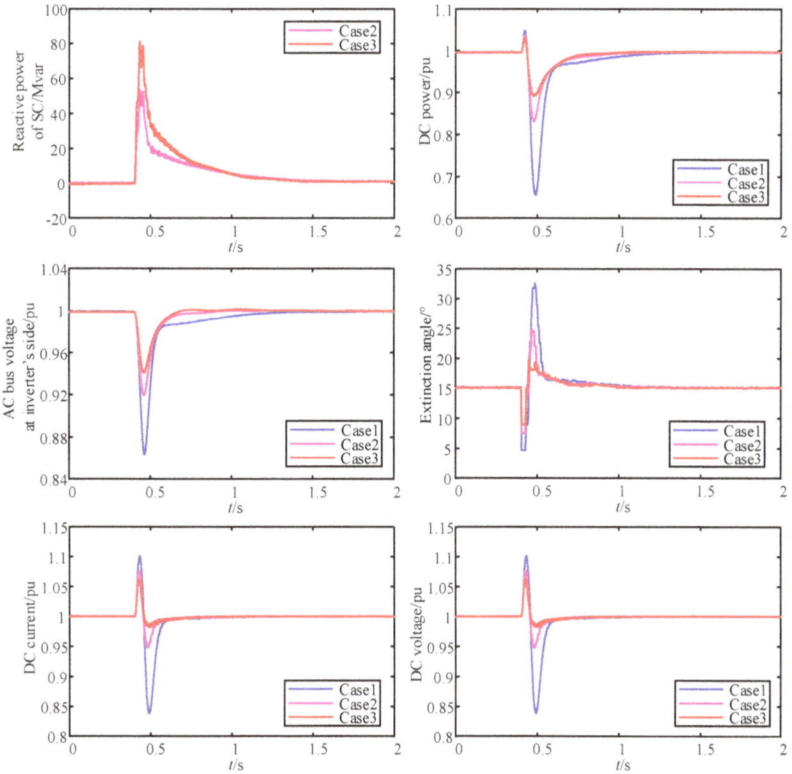

Figure 4. Feature comparison under the three-phase fault.

The comparison of the dynamic responses of different cases in Figure 4, showed that the DC voltage, DC current, DC power, AC bus voltage, γ and the reactive power of SCs all dropped or rose in varying degrees when the fault occured. Among them, the change range of Case 1 was the largest, on the contrary, it was the smallest in Case 3. Compared with Case 1, the AC bus voltage at the inverter side was supported and the rise of the DC current was restrained in Case 2 and Case 3 because of the reactive power compensation of the SCs, which reduced the drop of γ and effectively restrained the occurrence of commutation failures. The reactive power sent by the SCs in Case 3 was larger than that of Case 2 at the moment of fault. Therefore, Case 3 had a better suppression effect on commutation failure.

4.2. Proposed Area Ratio of Commutation Failure Probability

Previous studies have shown that different numbers of SCs can improve the ability of the LCC-HVDC system to resist commutation failure, but the method for quantitatively evaluating the suppression effect of SCs needs further study.

A single- or three-phase fault occurring at the AC bus of the inverter side may cause a commutation failure to occur in the LCC-HVDC system. The occurrence of a commutation failure is not only sensitive to the severity of the AC fault, but also to the time-instant of when the fault occurs during one AC cycle. Thus, the commutation failure probability index (CFPI) in [11] is used to assess the chances of commutation failure occurrence in the LCC-HVDC system during one cyclic frequency.

To obtain the commutation failure probability curve, different fault levels under specific fault times were considered. A single- or three-phase fault with inductance was applied at the AC busbar and the fault level was varied by changing the inductance value. A simultaneous fault level in the range of 100 fault points was applied while changing the fault time within one AC cycle duration (0.02 s). The multiple run mechanism in the PSCAD/EMTDC examined the occurrence of commutation failure and enlisted all related data. The probability of commutation failure for a specific fault was then evaluated by taking the ratio of faults that could cause commutation failure to the total equivalent fault points in one AC cycle. Multiple simulations were carried out to evaluate the probability for each fault level and to obtain the commutation failure probability curve.

However, when the CFPI was used to quantitatively evaluate the effect of the SC on the commutation failure of the LCC-HVDC system, the evaluation results were only based on a specific fault. It could not fully and intuitively reflect the effect of the SC on the commutation failure of the LCC-HVDC system in a wider fault range and thus could not provide a more comprehensive evaluation for the commutation failure mitigation effect. By comparing the curves of the commutation failure probability before and after the implementation of the SC at the inverter side, it was found that the area surrounded by the curve of the commutation failure probability with the SC was smaller than that without the SC within the same fault range. Considering this, an area ratio of commutation failure probability that could cope with the shortcomings of CFPI was proposed. The flow-chart for calculating the area ratio of commutation failure probability is shown in Figure 5.

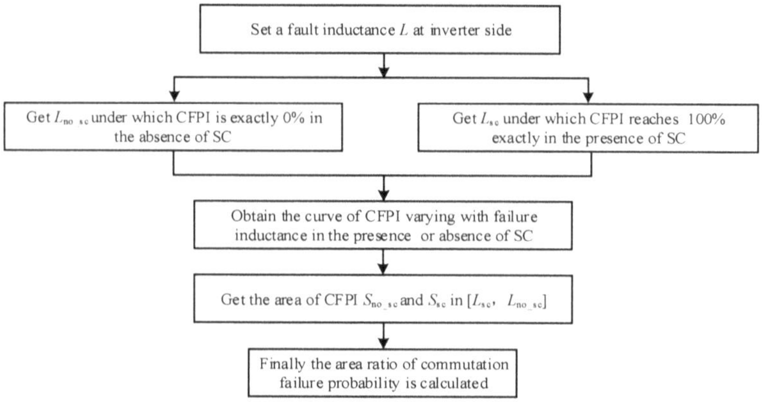

Figure 5. Flow-chart of the area ratio of the commutation failure probability computing method.

The detailed steps for calculating the area ratio of commutation failure probability are as follows:

(1) At the inverter end, the fault with a setting inductance value was applied to the AC busbar and the initial fault time was set with the fault clearance duration.

(2) When no SC was considered, the commutation failure probability of the LCC-HVDC system under a different fault inductance L was obtained. The fault inductance value was defined as $L_{\text{no\_sc}}$, under which the commutation failure probability of the LCC-HVDC system was exactly 0 %.

(3) Considering that SC was linked at the inverter end, the commutation failure probability of the LCC-HVDC system under a different fault inductance L was obtained. The fault inductance value was obtained as L_{sc}, under which the commutation failure probability of the LCC-HVDC reached 100%. (When there were multiple SCs, the L_{sc} that could take in the maximum number of SCs was chosen).

(4) With the fault inductance L as abscissa and the commutation failure probability as ordinate, the curve of commutation failure probability with and without SCs was formed as graphically presented in Figure 6. The area between the commutation failure probability curve and horizontal axis $[L_{\text{sc}}, L_{\text{no\_sc}}]$ was calculated. Finally, the area ratio of commutation failure probability was computed by taking the ratio of area of commutation failure probability without SC ($S_{\text{no\_sc}}$) to the area of commutation failure probability with SC (S_{sc}), as analytically expressed in (1).

$$Area\ Ratio = \frac{S_{\text{sc}}}{S_{\text{no\_sc}}} \tag{1}$$

It was clear from Figure 6 that for the same fault level, the commutation failure probability of LCC-HVDC with the SC was less than that of LCC-HVDC without the SC. Therefore, the calculated area-ratio of commutation failure probability should be less than 1. A smaller value of area ratio of commutation failure probability indicated a better effect of the SC in terms of suppressing the commutation failure of the LCC-HVDC system.

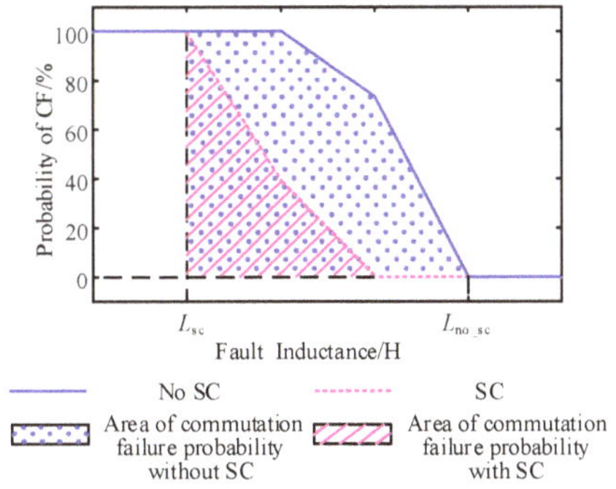

Figure 6. Schematic presentation of area of commutation failure probability with and without a SC.

4.3. Application and Validation of the Area Ratio of Commutation Failure Probability

In order to evaluate the suppression effect of the SC on the commutation failure of the LCC-HVDC system within a fault range, and to verify the proposed area ratio of commutation failure probability index under single-phase and three-phase inductive faults, the simulations were conducted when considering different numbers of SCs and different short circuit ratios (SCRs) at the inverter side, based on the electromagnetic transient model of the LCC-HVDC system with SCs.

The area of commutation failure probability in the three cases mentioned above were calculated using complex trapezoidal formula as in (2) and (3):

$$S = \frac{h}{2}[f(L_{\text{no\_sc}}) + f(L_{\text{sc}})] + h\sum_{k=1}^{n-1} f(L_k) \tag{2}$$

$$L_k = L_{sc} + k * h; (k = 1, 2 \ldots\ldots n - 1) \tag{3}$$

where,

h: step change of fault inductance in the simulation;
L_k: fault inductance in $[L_{sc}, L_{no\_sc}]$ interval;
$f(L_k)$: the commutation failure probability corresponding to the fault inductance L_k.
S: the area of commutation failure probability;
n: the total number of fault inductance level in $[L_{sc}, L_{no\_sc}]$;

The LCC-HVDC system was assumed to be operating at a rated nominal value (DC power = 1 p.u., DC voltage =1 p.u.). The short circuit ratios (SCRs) of the AC system at the rectifier and the inverter side of the LCC-HVDC system were both set at 2.5. The curve for the commutation failure probability of the LCC-HVDC system considering the three cases mentioned above during the single-phase inductive fault with changing inductance level is depicted in Figure 7. It is obvious from Figure 7 that in Case 3 with two SCs at the inverter side, the commutation failure probability was 100% for the single-phase fault with an inductance level of 0.25 H. Whereas in Case 2, the commutation failure probability was 100% for the fault inductance level of 0.35 H. Considering Case 1 without the SC, the commutation failure probability was 0% for the single-phase inductive fault with a 0.97 H inductance value. Using these values, the area ratio of commutation failure probability for the three cases under single-phase inductive fault was calculated using Equations (1)–(3).

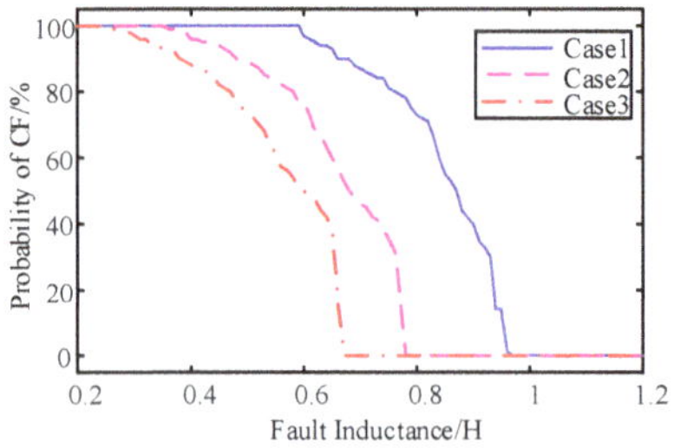

Figure 7. The probability of commutation failure (CF) in different cases under single phase to ground fault.

In order to indicate the severity of a single phase to ground fault, the minimum AC voltage RMS values at LCC inverter side (without a SC) during the fault period, at different fault inductances, are summarized in Table 2.

Table 2. The minimum AC voltage RMS values during the single phase to ground fault period.

Fault Inductance/H	Minimum AC Voltage/p.u.
0.2	0.868
0.3	0.904
0.4	0.920
0.5	0.934
0.6	0.943
0.7	0.949
0.8	0.953
0.9	0.956
1.0	0.960

Figure 8 shows the curve built for the commutation failure probability of the LCC-HVDC system under the three phase-ground inductive faults with an inductance range between 0.7 H and 1.4 H while assuming the three cases. It could be depicted that with two SCs, the commutation failure probability was 100 % under a fault inductance level of 0.78 H. The commutation failure probability with one SC (Case 2) was 100% for a fault inductance value of 0.91 H, whereas the commutation failure probability without the SC was 0% for a fault inductance level of 1.28 H. Considering these values in the three different cases, the area ratio of the commutation failure probability of the LCC-HVDC system under three-phase fault was calculated by applying Equations (1)–(3).

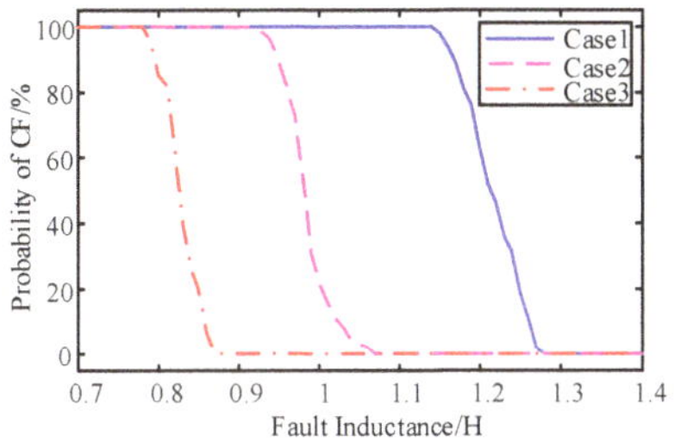

Figure 8. The probability of commutation failure (CF) in different cases under three phase fault.

In order to indicate the severity of the three phases to ground fault, the minimum AC voltage RMS values at the LCC inverter side (without a SC) during the fault period, at different fault inductances, are summarized in Table 3.

Table 3. The minimum AC voltage RMS values during the three phase to ground fault periods.

Fault Inductance/H	Minimum AC Voltage/p.u.
0.7	0.863
0.8	0.914
0.9	0.925
1.0	0.933
1.1	0.940
1.2	0.951

Table 4 summarizes the area of commutation failure probability and area ratio of commutation failure probability of the LCC-HVDC system with and without the SC under single-phase and three-phase inductive faults.

Table 4. Values of area of commutation failure probability and area ratio of commutation failure probability under different types of faults.

Index / Fault	Area of Commutation Failure Probability			Area Ratio of Commutation Failure Probability	
	Case 1 (no SC)	Case 2 (1 SC)	Case 3 (2 SCs)	Case 2 (1 SC)	Case 3 (2 SCs)
Single phase	0.5949	0.4128	0.3117	0.6939	0.5240
Three phase	0.4348	0.2038	0.0470	0.4687	0.1081

Table 2 shows that when a single-phase fault was applied to the AC bus at the inverter side, the area of commutation failure probability and the area ratio of commutation failure probability both decreased continuously with the increasing number of SCs. The area ratio of commutation failure probability was 0.6939 and 0.5240 with one SC (Case 2) and two SCs (Case 3), respectively, which indicated an improvement in Case 3 as the area ratio of commutation failure probability was reduced by 0.1699 compared to Case 2. Considering the three-phase fault, the area ratio of commutation failure probability of the LCC-HVDC system with one SC at the inverter side was 0.4687. With two SCs linked to the LCC-HVDC system (Case 3), the area ratio of commutation failure probability was 0.1081, which by comparing with Case 2 revealed that an increasing number of SCs could effectively suppress the commutation failure under the three-phase fault. Thus it could be clearly stated that the proposed area ratio of commutation failure probability could comprehensively evaluate the effect of the SCs on the commutation failure of the LCC-HVDC system under various fault types. In addition, the proposed index took into account various conditions of fault severities within a wider fault range, and was not limited to a specific fault. Comparing with existing indices such as the CFPI, the proposed index could quantitatively evaluate the impact of the SCs in suppressing the commutation failure of the LCC-HVDC system in a comprehensive and easy way.

Under different system intensities indicated by the short circuit ratio (SCR), the effect of the SCs on the ability of the LCC-HVDC system to resist commutation failure in a certain fault range varied. Therefore, the single-phase to ground and the three-phase faults were applied respectively at the inverter bus, then the area ratio of commutation failure probability was used to measure the suppression effect of the SCs on commutation failure under different SCRs to obtain the results shown in Figures 9 and 10, respectively.

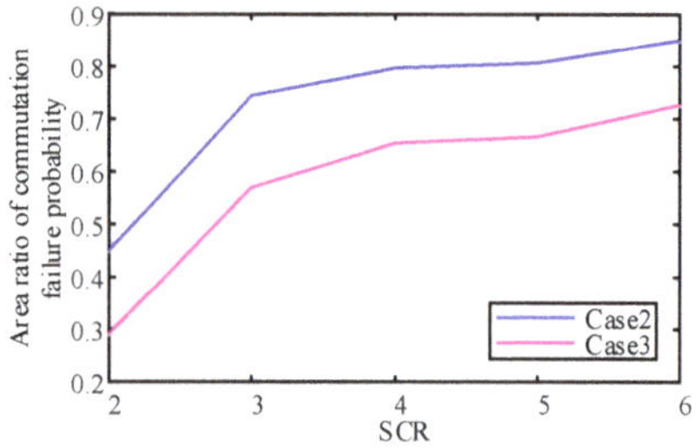

Figure 9. Area ratio of commutation failure probability in different cases under single phase to ground fault.

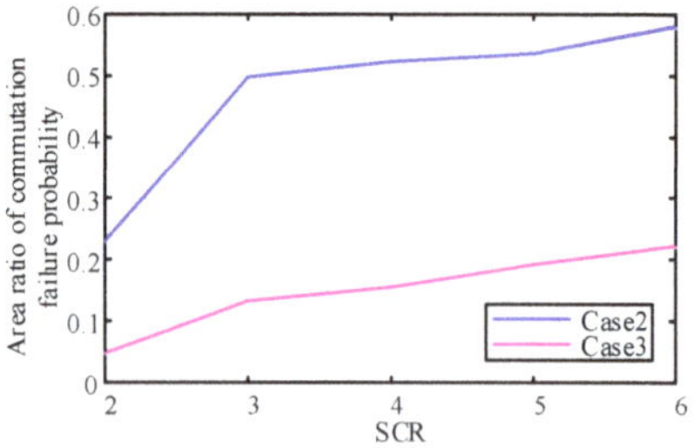

Figure 10. Area ratio of commutation failure probability in different cases under three phase fault.

It could be seen from Figure 9 that when a single-phase fault occurs at the AC bus of the inverter side, the area ratio of commutation failure probability in Case 2 and Case 3 were 0.4487 and 0.2893 respectively when SCR equalled 2; the area ratio of commutation failure probability in Case 2 and Case 3 were 0.8481 and 0.7246, respectively, when SCR equalled 6. It could be seen from Figure 10 that when the three-phase fault occured at the AC bus of the inverter side, the area ratio of commutation failure probability in Case 2 and Case 3 were 0.2305 and 0.0465 respectively when SCR equalled 2, the area ratio of commutation failure probability in Case 2 and Case 3 were 0.5786 and 0.2207, respectively, when SCR equalled 6. Regardless of whether it was under the single phase to ground or three-phase fault, the area ratio of the commutation failure probability of Case 3 was smaller than that of Case 2 under the same SCR, which indicated that an increasing number of SCs could improve the ability of the LCC-HVDC system to suppress the commutation failure. The area ratio of commutation failure probability in Case 3 and Case 2 increased with an increase in the short-circuit ratio, which showed that with the increase of system strength, the enhancement effect of the anti-commutation failure ability with SCs was weakened.

5. Conclusions

In this paper, based on the CIGRE benchmark, an electromagnetic transient model of a LCC-HVDC system with a synchronous condenser (SC) was developed in PSCAD/EMTDC. The area ratio of commutation failure probability was proposed, and the effect of the SC on suppressing the commutation failure of the LCC-HVDC system was quantitatively evaluated. The following conclusions were drawn:

The introduced area of commutation failure probability was visually intuitive. The proposed area ratio of commutation failure probability could comprehensively reflect the overall improvement effect of the SC on the commutation failure of the LCC-HVDC system, while considering the severity of different fault conditions. Making the assessment results more comprehensive could help in coping with the inadequacies (shortcomings) of the existing commutation failure probability index that was limited to evaluating only under specific faults.

Putting the SC at the inverter end of the LCC-HVDC system could enhance the support capability of the AC system at the receiving end due to the shunting effect of transient reactance even under a steady-state. Increasing the number of SC units, i.e., increasing the Mvar capacity of the SC to be capable of providing a maximum surplus of reactive power during transient conditions could significantly improve the commutation failure susceptibility of the LCC-HVDC system, especially in weak AC systems.

The proposed index could provide a theoretical basis for the capacity allocation of SCs in real world LCC-HVDC projects. Moreover, this index could also be used for evaluating the impact of other types of dynamic reactive power compensators on the commutation failure probability of the LCC-HVDC system during various AC disturbances.

Author Contributions: Conceptualization, J.S. and C.G.; Writing—Original Draft Preparation, J.S.; Writing—Review and Editing, J.S. and A.U.R.; Supervision, C.G. and C.Z.

Funding: This research was funded by National Natural Science Foundation of China (NSFC), grant number 51877077.

Conflicts of Interest: The authors declare no conflicts of interest.

References

1. Guo, C.; Li, C.; Zhao, C.; Ni, X.; Zha, K.; Xu, W. An Evolutional Line-Commutated Converter Integrated With Thyristor-Based Full-Bridge Module to Mitigate the Commutation Failure. *IEEE Trans. Power Electron.* **2017**, *32*, 967–976. [CrossRef]
2. Wei, Z.; Yuan, Y.; Lei, X.; Wang, H.; Sun, G.; Sun, Y. Direct-Current Predictive Control Strategy for Inhibiting Commutation Failure in HVDC Converter. *IEEE Trans. Power Syst.* **2014**, *29*, 2409–2417. [CrossRef]
3. Thio, C.V.; Davies, J.B.; Kent, K.L. Commutation Failures in HVDC Transmission Systems. *IEEE Trans. Power Deliv.* **1996**, *11*, 946–953. [CrossRef]
4. Sun, Y.Z.; Peng, L.; Ma, F.; Li, G.J.; Lv, P.F. Design a Fuzzy Controller to Minimize the Effect of HVDC Commutation Failure on Power System. *IEEE Trans. Power Syst.* **2008**, *23*, 100–107. [CrossRef]
5. Teleke, S.; Abdulahovic, T.; Thiringer, T.; Svensson, J. Dynamic Performance Comparison of Synchronous Condenser and SVC. *IEEE Trans. Power Deliv.* **2008**, *23*, 1606–1612. [CrossRef]
6. IEEE Power & Energy Society. *IEEE Guide for Planning DC Links Terminating at AC Locations Having Low Short-Circuit Capacities*; IEEE Standard Board: New York, NY, USA, 1997; pp. 1–216.
7. Zhou, Y.; Wu, H.; Wei, W.; Song, Y.; Deng, H. Optimal Allocation of Dynamic Var Sources for Reducing the Probability of Commutation Failure Occurrence in the Receiving-end Systems. *IEEE Trans. Power Deliv.* **2019**, *34*, 324–333. [CrossRef]
8. Xiao, H.; Li, Y.; Zhu, J.; Duan, X. Efficient approach to quantify commutation failure immunity levels in multi-infeed HVDC systems. *IET Gener. Transm. Dis.* **2016**, *10*, 1032–1038. [CrossRef]
9. Shao, Y.; Tang, Y. Fast Evaluation of Commutation Failure Risk in Multi-Infeed HVDC Systems. *IEEE Trans. Power Syst.* **2018**, *33*, 646–653. [CrossRef]

10. Yang, H.; Cai, Z.; Li, X.; Yu, C. Assessment of commutation failure in HVDC systems considering spatial-temporal discreteness of AC system faults. *J. Mod. Power Syst. Clean Energy* **2018**, *6*, 1055–1065. [CrossRef]
11. Rahimi, E.; Gole, A.M.; Davies, J.B.; Fernando, I.T.; Kent, K.L. Commutation failure analysis in multi-infeed HVDC systems. *IEEE Trans. Power Deliv.* **2011**, *26*, 378–384. [CrossRef]
12. Szechtman, M.; Wess, T.; Thio, C.V. A benchmark model for HVDC system studies. In Proceedings of the International Conference on AC and DC Power Transmission, London, UK, 17–20 September 1991; pp. 374–378.

MDPI
St. Alban-Anlage 66
4052 Basel
Switzerland
Tel. +41 61 683 77 34
Fax +41 61 302 89 18
www.mdpi.com

Applied Sciences Editorial Office
E-mail: applsci@mdpi.com
www.mdpi.com/journal/applsci

www.ingramcontent.com/pod-product-compliance
Lightning Source LLC
Chambersburg PA
CBHW051859210326
41597CB00033B/5952
* 9 7 8 3 0 3 9 2 1 7 6 2 5 *